AF361865

Application of Renewable Energy in Production and Supply Chain Management

Application of Renewable Energy in Production and Supply Chain Management

Special Issue Editor

Biswajit Sarkar

MDPI • Basel • Beijing • Wuhan • Barcelona • Belgrade • Manchester • Tokyo • Cluj • Tianjin

Special Issue Editor
Biswajit Sarkar
Yonsei University
Korea

Editorial Office
MDPI
St. Alban-Anlage 66
4052 Basel, Switzerland

This is a reprint of articles from the Special Issue published online in the open access journal *Energies* (ISSN 1996-1073) (available at: https://www.mdpi.com/journal/energies/special_issues/renewable_energy_production_supply_chain_management).

For citation purposes, cite each article independently as indicated on the article page online and as indicated below:

LastName, A.A.; LastName, B.B.; LastName, C.C. Article Title. *Journal Name* **Year**, *Article Number*, Page Range.

ISBN 978-3-03928-672-0 (Pbk)
ISBN 978-3-03928-673-7 (PDF)

Contents

About the Special Issue Editor

Biswajit Sarkar is currently Associate Professor at the Department of Industrial Engineering, Yonsei University, South Korea. He has completed his B.S. and M.S. in Applied Mathematics in 2002 and 2004, respectively, at Jadavpur University, India. He was awarded his Master of Philosophy for the application of Boolean polynomials from the Annamalai University, India, in 2008; Doctor of Philosophy in Operations Research from Jadavpur University, India, in 2010; and conducted his postdoctorate at the Pusan National University, South Korea (2012–2013). He has carried out teaching and research activities at various universities, including Hanyang University, South Korea (2014–2019), Vidyasagar University, India (2010–2014), and Darjeeling Government College, India (2009–2010). Under his supervision, fifteen students have been awarded their PhDs and three students their master's degree. Since 2010, he has published 166 journal articles in reputed journals such as Applied Mathematics and Industrial Engineering, and he has published one book. He serves on the Editorial Board of various reputed journals, including International Journals of Applied Mathematics and Industrial Engineering. He is the Topic Editor of the SCIE-indexed journal Energies and the Section Editor of the Scopus-indexed journal Inventions. He has served as Guest Editor of three Special Issues of the two SCIE-indexed journals Mathematics and Energies. He is a member of several learned societies. He was awarded Best Research Paper at an international conference in South Korea in 2014. He has been an Invited Speaker at numerous international conferences in addition to chairing sessions. He has been awarded a bronze medal from the Hanyang University in 2016 in recognition of his achievements. He is the recipient of the Bharat Vikash Award as a young scientist from India in 2016. He has received the International Award from Korean Institute of Industrial Engineers in 2017 at KAIST, Daejon, South Korea. He is the recipient of the Hanyang University Academic Award as one of the most productive researchers both in 2017 and 2018.

Preface to "Application of Renewable Energy in Production and Supply Chain Management"

The use of energy increases day by day with the increasing use of advanced technologies. Advanced technologies require more energy to run systems in addition to consuming basic traditional energies. As traditional energy resources are also currently the most heavily used energy resource for human beings, human civilization is facing an energy crisis due to the random use of energy by advanced machines. The issue is that the human civilization cannot survive without energy. Recent research has attempted to find ways to maximize the use of renewable energy with minimal use of traditional energy. In the industry, the production sector uses the maximum possible energy throughout the entire process of production and transportation. This book examines how advanced machines consume renewable energies in carrying out their various purposes by providing a glimpse into the recent research efforts that investigate the use of energy in the field of production, inventory, and supply chain management.

Readers can explore the different ways in which energy is utilized in the areas of production, inventory, and supply chains. They can experience how much energy is required for a particular item and the involved energy cost. This collection of articles provides new ideas and strategies for the use of different energies and their impact on the economy, society, and environment, and broadly deal with either production or supply chain management.

Thanks to all the contributors to the Special Issue "Application of Renewable Energy in Production and Supply Chain Management" of the SCIE indexed journal Energies. All of the ideas, results, and methods described in your research articles contribute to enriching the literature on energy.

Biswajit Sarkar
Special Issue Editor

Article

Effect of Energy and Failure Rate in a Multi-Item Smart Production System

Mitali Sarkar, Biswajit Sarkar * and Muhammad Waqas Iqbal

Department of Industrial & Management Engineering, Hanyang University, Ansan, Gyeonggi-do 155 88, Korea; mitalisarkar.ms@gmail.com (M.S.); waqastextilion@gmail.com (M.W.I.)
* Correspondence: bsbiswajitsarkar@gmail.com; Tel.: +82-10-7498-1981

Received: 6 October 2018; Accepted: 23 October 2018; Published: 30 October 2018

Abstract: To form a smart production system, the effect of energy and machines' failure rate plays an important role. The main issue is to make a smart production system for complex products that the system may produce several defective items during a long-run production process with an unusual amount of energy consumption. The aim of the model is to obtain the optimum amount of smart lot, the production rate, and the failure rate under the effect of energy. This study contains a multi-item economic imperfect production lot size energy model considering a failure rate as a system design variable under a budget and a space constraint. The model assumes an inspection cost to ensure product's quality under perfect energy consumption. Failure rate and smart production rate dependent development cost under energy consumption are considered, i.e., lower values of failure rate give higher values of development cost and vice versa under the effect of proper utilization of energy. The manufacturing system moves from in-control state to out-of-control state at a random time. The theory of nonlinear optimization (Kuhn–Tucker method) is employed to solve the model. There is a lemma to obtain the global optimal solution for the model. Two numerical examples, graphical representations, and sensitivity analysis of key parameters are given to illustrate the model.

Keywords: energy; multi-item smart production; system reliability; failure rate; variable development cost

1. Introduction

During a long-run production, a common phenomenon is the production of defective items even though the production is considered under a smart manufacturing system under the consideration of proper energy consumption. The rate of production of defective items may be of two types: constant defective rate and random defective rate. In constant defective rate, the total number of defective items are fixed, whereas in random defective rate, the number of defective items varies based on several conditions of the production system. In reality, both defective rates are available constant defective rate (see for reference [1]) and random defective rate [2]. Until now, no author considered random defective rate for multi-item smart production under energy consideration with budget and space constraints. Though the major contribution is the concept of failure rate of a smart production system under energy consideration being introduced with the random time movement from in-control state to out-of-control state. The failure rate is defined as the total number of failures divided by total number of working hours. The failure rate of a smart production system is considered as a system reliability indicator because a lower failure rate indicates more reliable systems and a higher failure rate indicates less reliable systems. Therefore, the proposed model gives a new direction of random defective rate with an indication of system reliability for multi-item smart production and energy consumption with budget and space constraints. Usually, for any long-run production system, it may contain production of both perfect and imperfect products. The imperfect products can either be discarded or can be reworked to make them perfect. This imperfection occurs when the system moves to *out-of-control* state, which is due to the factors such as machine breakdown, program inaccuracy, machine operator's inefficiency, and defective raw material supply. Several researchers have proposed inventory and production models with imperfect production systems. Kim and Hog [3] extended the Rossenblat and Lee [4] model within imperfect production systems by considering deteriorating production processes to obtain optimal production run length. They introduced the concept of system movement to *out-of-control* state from *in-control* state and producing defective items with three different deteriorating processes: constant, linearly increasing and exponentially increasing. Giri and Dohi [5] considered a random time of machine failure in an imperfect production system, where machine failure time and preventive time are random variables assuming stochastic machine breakdown and repair. They considered a net present value (NPV) approach for exact financial implications of the lot sizing to develop the EMQ model.

Sana et al. [6] extended the concept of an imperfect production system to introduce a new research dimension by considering a reduced selling-price for imperfect products. Reworking of the imperfect production items to make them as good as perfect quality items was introduced by Chiu et al. [7] and they proposed a model in which a portion of imperfect quality items is discarded, while the other portion is reworked by spending some costs. They optimized the finite production rate considering scrap production, reworking, and stochastic machine breakdown. The main research gap in this literature up to now is that no author utilized the concept of energy consumption and corresponding cost within any smart production system. González et al. [8] developed a model on turbomachinery components which are using for grinding flank tools. Egea et al. [9] implemented a short-cut method to measure the available energy in a required load capacity of a forging machine. They estimated the total energy during the friction of two screws.

Sarkar [10] developed an inventory model for retailers with a stock-depended demand and delay-in-payments considering that the replenished items are not all perfect presuming that the production system is imperfect and the inventory is replenished at a finite rate. An important managerial insight was added by Sarkar [11] by introducing a time dependent rate of product deterioration in an inventory system, where an inventory replenishment rate is finite and the customer is offered quantity discounts to attract a large order size in order to maximize the profit. Production of imperfect items depends upon the system reliability. The greater the investment in system development to increase its reliability, the lesser the production of imperfect items will be. System reliability-dependent imperfect production was discussed by Sarkar [12] for an inflationary economic manufacturing quantity (EMQ) system, where demand depends upon the product price and advertisement. Chakraborty and Giri [13] modeled an imperfect production system, where system shifts to *out-of-control* state during preventive maintenance and, during the state, some imperfect items are produced, which are inspected and reworked at the end of the production run. They also assumed that some of the reworked items cannot be repaired.

An economic production quantity model with random defective rate of imperfect items' production was investigated by Sarkar et al. [14] with rework process and planned backorders. They considered three different distribution density functions to calculate the rate of defective items and compared the results. Sarkar and Saren [15] studied deteriorating/imperfect production process, which randomly moves to an *out-of-control* state. They suggested that lot inspection policy should be adopted rather than full inspection policy to reduce the inventory costs. They also considered state of quality inspectors, who may falsely choose imperfect items as perfect and vice versa, which are designated as Type 1 and Type 2 errors. They also considered warranty policy over fixed time periods.

Pasandideh et al. [16] developed an inventory model for a multi-item single-machine lot size system with imperfect items' production. Those imperfect items are further classified on the basis of their failure severity for reworking and scrap. They considered that product shortages are backlogged, in order to make it more realistic. Purohit et al. [17] conducted a comprehensive detailed analysis of a lot size problem, an inventory control system for non-stationary stochastic demand considering constraints of carbon emissions and cycle service level using carbon cap-and-trade regulatory mechanism. They generalized the study on effects of emission parameters and properties of product as well as the performance on supply chain. Due to involvement of labor in production and considering their influence on production of defective items, it is considered important to invest in personnel training according to the adopted system. Sana [18] investigated with production of defective items and developed an economic production lot size model with the environment of production system when it moves to *out-of-control* state. Cárdenas-Barrón et al. [19] studies on optimal inventory with corrections and complements. Tiwari et al. [20,21] developed two models on deteriorating and partial backlogging.

Limited storage capacity for the inventory warehouse is now becoming a critical issue due to increasing costs of the storage facilities. This constraint is being considered by many researchers in situations where bulk production is being done. Huang et al. [22] developed an inventory model and investigated the optimal retailer's lot sizing policy under partially permissible delay-in-payments and space constraints. They considered extra cost payment for rental warehouse, when the capacity of the existing warehouse is full. Pasandideh and Niaki [23] developed a nonlinear integer programming model to solve an inventory model considering multi-items with space limitations. They found the optimal solution of the model within the available warehouse space by adding a space constraint. Hafshejani et al. [24] solved a multi-stage inventory model with a nonlinear cost function and space constraint through a genetic algorithm. Mahapatra et al. [25] introduced an inventory model with demand and reliability dependent unit production cost under limited space availability. They supposed that available space is limited with fuzzy variable and solved the storage space goal using an intuitionistic fuzzy optimization technique.

The manufacturers have a limited budget and resources based on a periodic budget plan. Hence, consideration of budget constraints into the model is more realistic. Some researchers already analyzed budget constrained situations. For example, Mohan et al. [26] developed an optimal replenishment policy for multi-item ordering under conditions of permissible delay-in-payments, a budget constraint, and permissible partial-payment at a penalty. Hou and Lin [27] calculated the optimal lot size and optimal capital investment in setup costs with a limited capital budget to minimize the expected total annual cost and to reduce the yield variability for random yield. Taleizadeh et al. [28] studied a multi-item production system considering imperfect items and reworking thereof. They included a service level and a budget constraint within the model and calculated the global minimum. Cárdenas-Barrón et al. [29] studied a production-inventory model in a just-in-time (JIT) system constrained with a maximum available budget and proposed a simple alternative heuristic algorithm to solve the model. Du et al. [30] and Todde et al. [31] developed models on energy analysis and energy consumption. Tomić and Schneider [32] explained the method of how energy can be recovered from waste by a closed-loop. Haraldsson and Johansson [33] studied on measures of different types of energy efficiency during production. Xu et al. [34] discussed the production of bio-fuel oil from pyrolysis products of plants. This model is extended in the direction of energy. See Table 1 for the contribution of the different authors.

Table 1. Research contribution by several authors.

Author(s)	System Reliability	Development Cost	Energy	Defective Rate	Constraint Type	Item Type
Rosenblatt and Lee [4]	NA	NA	NA	Constant	NA	Single
Giri and Dohi [5]	NA	NA	NA	Random	NA	Single
Sana et al. [6]	NA	NA	NA	Random	NA	Single
Sarkar [12]	Variable	Variable	NA	Random	NA	Single
Sarkar and Saren [16]	NA	NA	NA	Random	NA	Single
Sana [18]	Variable	Variable	NA	Random	NA	Single
Taleizadeh et al. [28]	NA	NA	NA	Constant	Budget	Multiple
Cárdenas-Barrón et al. [29]	NA	NA	NA	NA	Budget	Multiple
This paper	Variable	Variable	Considered	Random	Budget & Space	Multiple

NA indicates that is not applicable for that paper.

2. Problem Definition, Notation, and Assumptions

In this section, problem definition, notation and assumptions are given.

2.1. Problem Definition

A multi-item smart production system under an amount of energy consumption is considered with random defective items. At random time τ_i, the system moves to *out-of-control* state from *in-control* state and produce defective items. The random time τ_i follows an exponential distribution (see, for instance, Sana [2]). To make the system more reliable under an appropriate consumption of energy, the development cost and unit production cost are assumed variable with respect to the failure rate of the system. The unit production cost also depends on the variable smart production rate. An inspection is considered to obtain the defective items as the system moves to *out-of-control* state. The defective items are reworked and transferred as perfect quality. The aim is to obtain maximum profit for a multi-item smart production system under the proper energy consumption by considering system failure rate and random defective rate (see Figure 1).

The following notation and assumptions are used to develop the model.

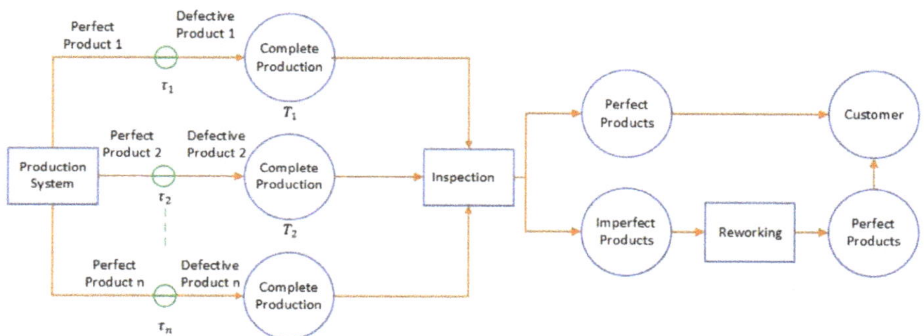

Figure 1. Process flow for multi-product production system

2.2. Assumptions

The following assumptions are considered to develop this model:

1. The model consists of a multi-item smart production system with variable production rate under the perfect consumption of energy and it is greater than the demand (D) such that there is no shortage.
2. The effect of energy is considered for the whole production system with an exact amount of energy consumption. The amount of energy consumption for holding inventory, inspection, rework, development, and tool/die cost is considered.
3. During long-run production, at any random time τ_i, the smart production system moves from *in-control* state to *out-of-control* state. The shifting time τ_i from *in-control* state to *out-of-control* state and the smart production rate follows a relation within them, where τ_i follows an exponential distribution by considering the failure rate η, which is a system design variable. If failure rate decreases, the system will be more reliable and, with the failure rate increasing, the system will be less reliable (see, for instance, Sarkar [12]).
4. To increase the system reliability, the unit smart production cost is assumed as a sum of development cost, material cost and tool/die cost, where the development cost of products depends on the failure rate (see, for instance, Sana [2]) and amount of energy consumption.
5. The model assumes multi-item smart production and there is a possibility of a space problem along with problem of total budget. The study considers space and budget constraints to solve

these types of issues such that the model becomes more realistic (see for instance [28,29]) under the efficient energy consumption.

6. In this model, although it is considered that the system will move to *out-of-control* state from *in-control* state after a certain time τ_i, production disruption is not considered here (see for instance Sarkar and Saren [15]).
7. The smart production system under energy consideration is considered for completely finished products. Work-in-process is not considered here.
8. Lead time is assumed as negligible.

3. Mathematical Model

This study contains a production-inventory model with a multi-item under energy consideration. The smart production continues from $t_i = 0$ to $t_i = t_{1i}$ for multi-item with a finite rate, where $t_i = Q_i/P_i$. The inventory piles up within the interval $[0, t_{1i}]$ and depletes within the interval $[t_{1i}, T]$ with demand D_i. The model considers that, after a random time τ_i, the system moves from *in-control* state to *out-of-control* state and produces imperfect products. See Figure 2 for the description of the production system.

The governing differential equation of the on-hand inventory is given by

$$\frac{dI_{1i}(t_i)}{dt_i} = P_i - D_i, \ 0 \le t_i \le t_{1i},$$

with initial condition $I_{1i}(0) = 0$, $i = 1, 2, ..., n$.
$$(1)$$

$$\frac{dI_{2i}(t_i)}{dt_i} = -D_i, \ t_{1i} \le t_i \le T,$$

with initial condition $I_{2i}(T) = 0$, $i = 1, 2, ..., n$.
$$(2)$$

The present state of inventories are given by

$$I_{1i}(t_i) = (P_i - D_i)t_i, \ 0 \le t_i \le t_{1i}, \ i = 1, 2, ..., n, \tag{3}$$

$$I_{2i}(t_i) = D_i(T - t_i), \ t_{1i} \le t_i \le T, \ i = 1, 2, ..., n. \tag{4}$$

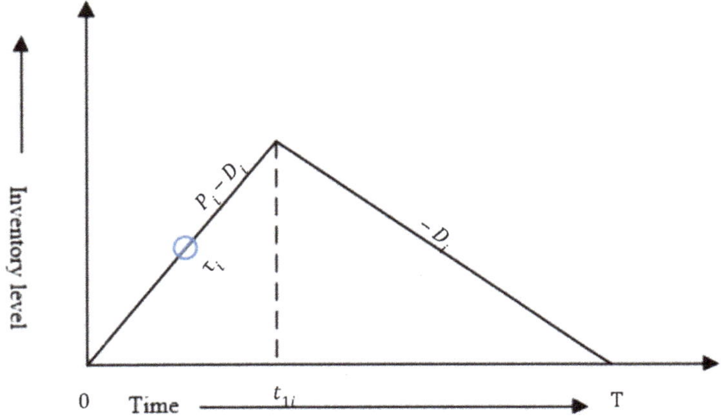

Figure 2. Economic production quantity model for multi-product systems.

The model now considers the following costs to calculate the profit of the smart production system.

3.1. Setup Cost (SC)

Setup cost plays a very important role for multi-item smart production systems as each item contains a different setup system with different energy consumption. Thus, the model assumes that the setup cost for *ith* item is considered as C_{si} per setup with C_{si}' as energy consumption cost per setup. Therefore, the average setup cost per unit cycle is

$$SC = \sum_{i=1}^{n} (C_{si} + C_{si}') \frac{D_i}{Q_i}.$$

3.2. Holding Cost (HC)

To calculate the holding cost for a smart multi-item production system, the average inventory for *ith* item has to calculate and, by taking summation over $i = 1$ to n, one can obtain the total inventory over the cycle length of the smart production system. Therefore, the total inventory divided by the cycle length of the production cycle gives the average inventory and per unit holding cost multiplied by the average inventory gives the average holding cost per cycle.

Hence, for calculating the total inventory, one has

$$\text{Inventory} = \sum_{i=1}^{n} \left[\int_0^{t_{1i}} I_{1i}(t_i) dt_i + \int_{t_{1i}}^{T} I_{2i}(t_i) dt_i \right]$$

$$= \sum_{i=1}^{n} \left[\int_0^{t_{1i}} (P_i - D_i) t_i dt_i + \int_{t_{1i}}^{T} D_i (T - t_i) dt_i \right].$$

As energy consumption is calculated with the appropriate costs, the holding cost for average inventory per unit time under the presence of cost for energy consumption is

$$HC = \sum_{i=1}^{n} \frac{(C_{hi} + C_{hi}') D_i}{2 Q_i} \left[\int_0^{t_{1i}} (P_i - D_i) t_i dt_i + \int_{t_{1i}}^{T} \left\{ (P_i - D_i) \frac{Q_i}{P_i} - D_i t_i \right\} dt_i \right]$$

$$= \sum_{i=1}^{n} \frac{(C_{hi} + C_{hi}') Q_i}{2} \left(1 - \frac{D_i}{P_i} \right).$$

3.3. Inspection Cost (IC)

During a long-run process, the smart production system may move to *out-of-control* state, thus an inspection of each product is necessary. By inspection, the industry can assure the good quality of products, which generally maintain the brand image of the industry. If inspection cost per unit is C_i and C_i' is the cost per unit for energy consumed due to inspection, then the inspection cost per unit cycle under energy consideration is

$$IC = \sum_{i=1}^{n} (C_i + C_i') Q_i \times \frac{D_i}{Q_i}$$

$$= \sum_{i=1}^{n} (C_i + C_i') D_i.$$

3.4. Rework Cost (RC)

After inspection of each product, those items, detected as defective, are considered for reworking to make them as if they are perfect. To calculate the rework cost, the number of defective items and the rate of defective items production are needed.

The rate of defective items $g(t_i, \tau_i, P_i)$ is considered (see, for instance, [2]) as

$$g(t_i, \tau_i, P_i) = \alpha P_i^\beta (t_i - \tau_i)^\gamma, \text{ where } \beta \geq 0, \gamma \geq 0 \text{ and } t_i \geq \tau_i. \tag{5}$$

There is a quality level of smart production defined by the management system of the smart production industry, below which a product will not remain qualitative. The items that do not qualify the requirements of quality are imperfect items and cannot be forwarded to customers before reworking. The production system produces defective from random time τ_i till time t_{1i}, which is the time for maximum inventory.

There is no imperfect items within the interval $[0, \tau_i]$ and all imperfect items produce within $[\tau_i, t_{1i}]$. Thus, number of imperfect items within the interval $[\tau_i, t_{1i}]$ is

$$\begin{aligned}
N &= P_i \int_{\tau_i}^{t_{1i}} \alpha P_i^\beta (t_i - \tau_i)^\gamma dt \\
&= \left(\frac{\alpha}{\gamma+1}\right) P_i^{\beta+1} (t_{1i} - \tau_i)^{\gamma+1}.
\end{aligned} \tag{6}$$

Therefore, the number of imperfect items within the full cycle is

$$N = \begin{cases} 0, & \text{if } \tau_i \geq t_{1i}, \\ \left(\frac{\alpha}{\gamma+1}\right) P_i^{\beta+1} (t_{1i} - \tau_i)^{\gamma+1}, & \text{if } \tau_i \leq t_{1i}, \end{cases}$$

where the random time τ_i follows the exponential distribution.

The distribution function of τ_i within the *out-of-control* state is considered as

$$G(\tau_i) = 1 - e^{-\eta \tau_i}, \tag{7}$$

where η is the failure rate, known as system design variable. The lower value of η indicates a higher value of system reliability. Now, to ensure the distribution function, it can be found easily

$$\int_0^\infty dG(\tau_i) = 1.$$

Generally, the rate of defective items' production cannot be determined. However, on the basis of previous data, an expected number of defective items' production can be calculated. We are adding those expected number of produced defective items to calculate the cost of imperfect products. Thus, the density function for the random time τ_i has to consider for calculation of the expected number of defective items within a full cycle. Hence, the expected number of imperfect items for the full cycle is

$$\begin{aligned}
E(N) &= \sum_{i=1}^n \left(\frac{\alpha}{\gamma+1}\right) P_i^{\beta+1} \int_0^{t_{1i}} (t_{1i} - \tau_i)^{\gamma+1} dG(\tau_i) \\
&= \sum_{i=1}^n \eta P_i^{\beta+1} \left(\frac{\alpha}{\gamma+1}\right) e^{\frac{-\eta Q_i}{P_i}} \psi\left(\eta, \frac{Q_i}{P_i}\right), \text{ as } t_{1i} = \frac{Q_i}{P_i}.
\end{aligned} \tag{8}$$

To change the status of defective products, the rework cost along with the cost for energy consumption during reworking is used to make them perfect as new. The rework cost per unit cycle (RC) is

$$RC = \sum_{i=1}^n (R_i + R_i') \frac{D_i}{Q_i} E(N).$$

3.5. Development Cost (DC)

To make the system more reliable, the failure rate, which in turn indicates the system reliability, is considered within the development cost of products. The labor cost and energy resource cost are included within it. Thus, the development cost per unit time is considered as

$$C_1(\eta) = M + Xe^{r\frac{\eta_{max}-\eta}{\eta-\eta_{min}}}.$$

(9)

3.6. Unit Production Cost (UPC)

Unit production cost is considered as the sum of raw material cost per product, development cost per product and tool/die cost. The unit production is directly proportional to the material cost as the increasing raw material cost indicates the increasing value of the unit production cost. It is also directly proportional to development cost and tool/die cost, as increasing the value of these costs results in more unit production cost. Unit production cost per unit time is assumed as

$$C_p(\eta, P_i) = \sum_{i=1}^{n} \left[C_m + \frac{C_1(\eta)}{P_i} + \alpha_1 P_i^{\delta} \right],$$

(10)

where C_m is the material cost per unit item, whose quality helps to make the system more reliable. $C_1(\eta)$ is the development cost which depends on failure rate η. With the increasing percentage of failure rate, the development cost increases, which indicates more reliable system as η, the failure rate, indicates the system reliability, and, when it decreases, development cost decreases. $\alpha P_i^{\delta} (\delta > 0)$ is the tool/die cost.

3.7. Expected Total Profit (ETP)

The expected total profit per unit cycle is ETP (Q_i, P_i, η) = Revenue $-$ HC $-$ SC $-$ IC $-$ RC

$$ETP(Q_i, P_i, \eta) = \sum_{i=1}^{n} \left[D_i(W_i - C_p) - \frac{(C_{hi} + C'_{hi})Q_i}{2} \left(1 - \frac{D_i}{P_i} \right) - (C_{si} + C'_{si})\frac{D_i}{Q_i} \right.$$

$$\left. - (C_i + C'_i)D_i - (R_i + R'_i) \left(\frac{D_i \alpha}{Q_i(\gamma + 1)} \right) P_i^{\beta+1} \eta e^{\frac{-\eta Q_i}{P_i}} \psi \left(\eta, \frac{Q_i}{P_i} \right) \right],$$

(11)

as $t_{2i} = \frac{(P_i - D_i)Q_i}{P_i D_i}$, $\left(\text{see Appendix A for the value of } \psi \left(\eta, \frac{Q_i}{P_i} \right)\right)$.

3.8. Constraints

In any business system, investment is not unlimited. With the available capital, a manufacturer can buy the plausible combinations of materials and services to satisfy the demand of its customer. Similarly, in an imperfect production system, only a specific percentage of budget can be allocated for inspection and reworking of imperfect items. There is a certain quality level, below which the threshold of the allocated budget is crossed and that imperfect item will not be reworked. This model considers a budget constraint and the managers define a specific quality level/threshold quality level to separate the imperfect products, which can be reworked or not chosen for reworking. Like budget, space is also a constraint in any type of production system. Excess inventory and space are used and trigger additional costs and thus the aims to eliminate excess space and inventory. For an imperfect production system, a limited space is allocated to store and rework the imperfect production.

Thus, considering budget and space constrains, the profit equation becomes

$$ETP(Q_i, P_i, \eta) = \sum_{i=1}^{n} \left[D_i(W_i - C_p) - \frac{(C_{hi} + C'_{hi})Q_i}{2} \left(1 - \frac{D_i}{P_i}\right) - (C_{si} + C'_{si})\frac{D_i}{Q_i} \right.$$

$$\left. - (C_i + C'_i)D_i - (R_i + R'_i)\left(\frac{D_i\alpha}{Q_i(\gamma + 1)}\right) P_i^{\beta+1}\eta e^{\frac{-\eta Q_i}{P_i}} \psi\left(\eta, \frac{Q_i}{P_i}\right) \right],$$

subject to

$$\sum_{i=1}^{n} \xi_i Q_i \leq A,$$

$$\sum_{i=1}^{n} \phi_i Q_i \leq B,$$

$$(12)$$

where the first term indicates revenues, the second term gives the holding cost and energy consumption cost due to holding products, the third term provides a setup cost and energy consumption of setup cost, the fourth term indicates inspection cost and energy utilization cost for inspection, the fifth cost is for reworking and the use of energy cost for reworking, and the next two terms are for space and budget constraints.

To obtain the maximum profit with respect to the optimum production quantity, production rate, and failure rate, the model has to solve with the best solution approach, which is described in the next section.

4. Solution Methodology

The profit function is highly nonlinear and it contains inequality constraints. Thus, the Kuhn–Tucker method is the best approach to solve this model.

Therefore, using the Kuhn–Tucker condition, the solution can be obtained as follows:

Lagrange equation of the above profit function is given by

$$L(Q_i, P_i, \eta, \lambda_1, \lambda_2) = \sum_{i=1}^{n} \left[D_i(W_i - C_p) - \frac{(C_{hi} + C'_{hi})Q_i}{2} \left(1 - \frac{D_i}{P_i}\right) - (C_{si} + C'_{si})\frac{D_i}{Q_i} \right.$$

$$- (C_i + C'_i)D_i - \frac{\eta \zeta_1 P_i^{\beta+1}}{Q_i} e^{\frac{-\eta Q_i}{P_i}} \psi\left(\eta, \frac{Q_i}{P_i}\right) + \lambda_1(\xi_i Q_i - A)$$

$$\left. + \lambda_2(\phi_i Q_i - B) \right],$$

where λ_1 and λ_2 are Lagrange multiplier and $\zeta_1 = \frac{\alpha R_i D_i}{\gamma + 1}$.

From the necessary condition of optimization of the Kuhn–Tucker method, one can obtain

$$\frac{\partial L}{\partial Q_i} = -\frac{(C_{hi} + C'_{hi})}{2}\left(1 - \frac{D_i}{P_i}\right) + \frac{(C_{si} + C'_{si})D_i}{Q_i^2} + \frac{\eta \zeta_1 P_i^{\beta+1}}{Q_i} e^{\frac{-\eta Q_i}{P_i}} \psi\left(\eta, \frac{Q_i}{P_i}\right)$$

$$+ \frac{\eta^2 \zeta_1 P_i^{\beta}}{Q_i} e^{\frac{-\eta Q_i}{P_i}} \psi\left(\eta, \frac{Q_i}{P_i}\right) - \frac{\eta \zeta_1 P_i^{\beta+1}}{Q_i} e^{\frac{-\eta Q_i}{P_i}} \frac{\partial \psi}{\partial Q_i} + \lambda_1 \xi_i + \lambda_2 \phi_i \geq 0,$$

$$(13)$$

$$\frac{\partial L}{\partial P_i} = \frac{D_i C_1(\eta)}{P_i^2} - D_i(\alpha_1 + \alpha'_1)\delta P_i^{\delta-1} - \frac{(C_{hi} + C'_{hi})D_i Q_i}{2P_i^2} - \frac{\eta \zeta_1 (\beta+1) P_i^{\beta}}{Q_i} e^{\frac{-\eta Q_i}{P_i}} \psi\left(\eta, \frac{Q_i}{P_i}\right)$$

$$- \frac{\eta^2 \zeta_1}{P_i^2} e^{\frac{-\eta Q_i}{P_i}} \psi\left(\eta, \frac{Q_i}{P_i}\right) - \frac{\eta \zeta_1 P_i^{\beta+1}}{Q_i} e^{\frac{-\eta Q_i}{P_i}} \frac{\partial \psi}{\partial P_i} \geq 0,$$

$$(14)$$

$$\frac{\partial L}{\partial \eta} = \frac{D_i N r(\eta_{max} - \eta_{min})}{P_i(\eta - \eta_{min})^2} e^{\frac{r(\eta_{max} - \eta)}{\eta - \eta_{min}}} - \frac{\zeta_1 P_i^{\beta+1}}{Q_i} e^{\frac{-\eta Q_i}{P_i}} \psi\left(\eta, \frac{Q_i}{P_i}\right)$$
$$+ \frac{\eta \zeta_1 P_i^{\beta}}{Q_i} e^{\frac{-\eta Q_i}{P_i}} \psi\left(\eta, \frac{Q_i}{P_i}\right) - \frac{\eta \zeta_1 P_i^{\beta+1}}{Q_i} e^{\frac{-\eta Q_i}{P_i}} \frac{\partial \psi}{\partial \eta} \geq 0. \tag{15}$$

From the Kuhn–Tucker condition, one can write

$$\frac{(C_{si} + C'_{si}) D_i}{Q_i^2} + \frac{\eta \zeta_1 P_i^{\beta+1}}{Q_i} e^{\frac{-\eta Q_i}{P_i}} \psi\left(\eta, \frac{Q_i}{P_i}\right) - \frac{(C_{hi} + C'_{hi})}{2}\left(1 - \frac{D_i}{P_i}\right)$$
$$+ \frac{\eta^2 \zeta_1 P_i^{\beta}}{Q_i} e^{\frac{-\eta Q_i}{P_i}} \psi\left(\eta, \frac{Q_i}{P_i}\right) - \frac{\eta \zeta_1 P_i^{\beta+1}}{Q_i} e^{\frac{-\eta Q_i}{P_i}} \frac{\partial \psi}{\partial Q_i} + \lambda_1 \xi_i + \lambda_2 \phi_i = 0, \tag{16}$$

$$\frac{D_i C_1(\eta)}{P_i^2} - D_i(\alpha_1 + \alpha'_1)\delta P_i^{\delta-1} - \frac{(C_{hi} + C'_{hi})D_i Q_i}{2P_i^2} - \frac{\eta \zeta_1(\beta+1)P_i^{\beta}}{Q_i} e^{\frac{-\eta Q_i}{P_i}} \psi\left(\eta, \frac{Q_i}{P_i}\right)$$
$$- \frac{\eta^2 \zeta_1}{P_i^2} e^{\frac{-\eta Q_i}{P_i}} \psi\left(\eta, \frac{Q_i}{P_i}\right) - \frac{\eta \zeta_1 P_i^{\beta+1}}{Q_i} e^{\frac{-\eta Q_i}{P_i}} \frac{\partial \psi}{\partial P_i} = 0, \tag{17}$$

$$\frac{D_i N r(\eta_{max} - \eta_{min})}{P_i(\eta - \eta_{min})^2} e^{\frac{r(\eta_{max} - \eta)}{\eta - \eta_{min}}} - \frac{\zeta_1 P_i^{\beta+1}}{Q_i} e^{\frac{-\eta Q_i}{P_i}} \psi\left(\eta, \frac{Q_i}{P_i}\right)$$
$$+ \frac{\eta \zeta_1 P_i^{\beta}}{Q_i} e^{\frac{-\eta Q_i}{P_i}} \psi\left(\eta, \frac{Q_i}{P_i}\right) - \frac{\eta \zeta_1 P_i^{\beta+1}}{Q_i} e^{\frac{-\eta Q_i}{P_i}} \frac{\partial \psi}{\partial \eta} = 0. \tag{18}$$

To show the global optimality of the profit function, the sufficient condition from the Kuhn–Tucker condition must be satisfied. To obtain the global optimal solution, a lemma is established as follows:

Lemma 1. *(i) If $\zeta_2 > 0$, $\zeta_2 \zeta_3 - (\zeta_5)^2 < 0$, $(\zeta_2 \zeta_3 \zeta_4 - \zeta_2 \zeta_7^2 - \zeta_5^2 \zeta_4 + 2\zeta_5 \zeta_6 \zeta_7 - \zeta_6^2 \zeta_3) > 0$ or (ii) $\zeta_2 < 0$, $\zeta_2 \zeta_3 - (\zeta_5)^2 > 0$, $(\zeta_2 \zeta_3 \zeta_4 - \zeta_2 \zeta_7^2 - \zeta_5^2 \zeta_4 + 2\zeta_5 \zeta_6 \zeta_7 - \zeta_6^2 \zeta_3) < 0$ then $L(Q_i, P_i, \eta, \lambda_1, \lambda_2)$ at (Q_i^*, P_i^*, η^*) will be maximum, when*

$$-\frac{(C_{hi} + C'_{hi})}{2}\left(1 - \frac{D_i}{P_i}\right) + \frac{(C_{si} + C'_{si})D_i}{(Q_i^*)^2} + \frac{\eta \zeta_1 P_i^{\beta+1}}{Q_i^*} e^{\frac{-\eta Q_i^*}{P_i}} \psi\left(\eta, \frac{Q_i^*}{P_i}\right) + \frac{\eta^2 \zeta_1 P_i^{\beta}}{Q_i^*} e^{\frac{-\eta Q_i^*}{P_i}} \psi\left(\eta, \frac{Q_i^*}{P_i}\right)$$
$$- \frac{\eta \zeta_1 P_i^{\beta+1}}{Q_i^*} e^{\frac{-\eta Q_i^*}{P_i}} \frac{\partial \psi}{\partial Q_i^*} + \lambda_1 \xi_i + \lambda_2 \phi_i = 0, \quad \frac{-(C_{hi} + C'_{hi})D_i Q_i}{2(P_i^*)^2} - \frac{\eta \zeta_1(\beta+1)(P_i^*)^{\beta}}{Q_i} e^{\frac{-\eta Q_i}{P_i^*}} \psi\left(\eta, \frac{Q_i}{P_i^*}\right)$$
$$- \frac{\eta^2 \zeta_1}{(P_i^*)^2} e^{\frac{-\eta Q_i}{P_i^*}} \psi\left(\eta, \frac{Q_i}{P_i^*}\right) - \frac{\eta \zeta_1 (P_i^*)^{\beta+1}}{Q_i} e^{\frac{-\eta Q_i}{P_i^*}} \frac{\partial \psi}{\partial P_i^*} = 0, \text{ and } - \frac{\zeta_1 P_i^{\beta+1}}{Q_i} e^{\frac{-\eta^* Q_i}{P_i}} \psi\left(\eta^*, \frac{Q_i}{P_i}\right)$$
$$+ \frac{\eta^* \zeta_1 P_i^{\beta}}{Q_i} e^{\frac{-\eta^* Q_i}{P_i}} \psi\left(\eta^*, \frac{Q_i}{P_i}\right) - \frac{\eta^* \zeta_1 P_i^{\beta+1}}{Q_i} e^{\frac{-\eta^* Q_i}{P_i}} \frac{\partial \psi}{\partial \eta^*} = 0.$$

Proof. To find the sufficient condition of the global optimality, taking 2nd order derivatives of Equations (13)–(15) with respect to Q_i, P_i, and η, it becomes

$$\frac{\partial^2 L}{\partial Q_i^2} = -\frac{2(C_{si} + C'_{si})D_i}{Q_i^3} - \frac{2\eta \zeta_1 P_i^{\beta+1}}{Q_i^3} e^{\frac{-\eta Q_i}{P_i}} \psi\left(\eta, \frac{Q_i}{P_i}\right) - \frac{2\eta^2 \zeta_1 P_i^{\beta}}{Q_i^2} e^{\frac{-\eta Q_i}{P_i}} \psi\left(\eta, \frac{Q_i}{P_i}\right)$$
$$+ \frac{2\eta \zeta_1 P_i^{\beta+1}}{Q_i^2} e^{\frac{-\eta Q_i}{P_i}} \frac{\partial \psi}{\partial Q_i} - \frac{\eta^3 \zeta_1 P_i^{\beta-1}}{Q_i} e^{\frac{-\eta Q_i}{P_i}} \psi\left(\eta, \frac{Q_i}{P_i}\right) + \frac{2\eta^2 \zeta_1 P_i^{\beta}}{Q_i} e^{\frac{-\eta Q_i}{P_i}} \frac{\partial \psi}{\partial Q_i}$$
$$- \frac{\eta \zeta_1 P_i^{\beta+1}}{Q_i} e^{\frac{-\eta Q_i}{P_i}} \frac{\partial^2 \psi}{\partial Q_i^2}$$
$$= \zeta_2 \text{ (say)},$$

$$\frac{\partial^2 L}{\partial P_i^2} = \frac{2(C_{hi} + C'_{hi})D_i}{P_i^3} - \frac{\eta\zeta_1\beta(\beta+1)P_i^{\beta-1}}{Q_i}e^{\frac{-\eta Q_i}{P_i}}\psi\left(\eta, \frac{Q_i}{P_i}\right) - \eta^2\zeta_1(\beta+1)P_i^{\beta+2}e^{\frac{-\eta Q_i}{P_i}}\psi\left(\eta, \frac{Q_i}{P_i}\right)$$

$$- \frac{2\eta^2\zeta_1}{P_i^3}e^{\frac{-\eta Q_i}{P_i}}\psi\left(\eta, \frac{Q_i}{P_i}\right) - \frac{\eta^3\zeta_1 Q_i}{P_i^4}e^{\frac{-\eta Q_i}{P_i}}\psi\left(\eta, \frac{Q_i}{P_i}\right) - \frac{\eta^2\zeta_1}{P_i^2}e^{\frac{-\eta Q_i}{P_i}}\frac{\partial\psi}{\partial P_i}$$

$$- \frac{\eta\zeta_1(\beta+1)P_i^{\beta}}{Q_i}e^{\frac{-\eta Q_i}{P_i}}\frac{\partial\psi}{\partial P_i} - \eta^2\zeta_1 P_i^{\beta-1}e^{\frac{-\eta Q_i}{P_i}}\frac{\partial\psi}{\partial P_i} - \frac{\eta\zeta_1 P_i^{\beta+1}}{Q_i}e^{\frac{-\eta Q_i}{P_i}}\frac{\partial^2\psi}{\partial P_i^2}$$

$$= \zeta_3 \text{ (say)},$$

and

$$\frac{\partial^2 L}{\partial\eta^2} = 2\zeta P_i^{\beta}e^{\frac{-\eta Q_i}{P_i}}\psi\left(\eta, \frac{Q_i}{P_i}\right) - 2\frac{\zeta_1 P_i^{\beta+1}}{Q_i}e^{\frac{-\eta Q_i}{P_i}}\frac{\partial\psi}{\partial\eta} - \eta\zeta_1 P_i^{\beta-1}Q_i e^{\frac{-\eta Q_i}{P_i}}\psi\left(\eta, \frac{Q_i}{P_i}\right)$$

$$+ 2\eta\zeta_1 P_i^{\beta}e^{\frac{-\eta Q_i}{P_i}}\frac{\partial\psi}{\partial\eta} - \frac{\eta\zeta_1 P_i^{\beta+1}}{Q_i}e^{\frac{-\eta Q_i}{P_i}}\frac{\partial^2\psi}{\partial\eta^2}$$

$$= \zeta_4 \text{ (say)}.$$

Now, taking derivatives of Equation (13) with respect to P_i and η, one can obtain

$$\frac{\partial^2 L}{\partial Q_i \partial P_i} = -\frac{(C_{hi} + C'_{hi})D_i}{2P_i^2} + \frac{\eta\zeta_1(\beta+1)P_i^{\beta}}{Q_i^2}e^{\frac{-\eta Q_i}{P_i}}\psi\left(\eta, \frac{Q_i}{P_i}\right) + \frac{\eta^2\zeta_1(\beta+1)P_i^{\beta-1}}{Q_i}e^{\frac{-\eta Q_i}{P_i}}\psi\left(\eta, \frac{Q_i}{P_i}\right)$$

$$- \eta\zeta_1(\beta+1)P_i^{\beta}e^{\frac{-\eta Q_i}{P_i}}\frac{\partial\psi}{\partial Q_i} + \frac{\eta^3\zeta_1}{P_i^3}e^{\frac{-\eta Q_i}{P_i}}\psi\left(\eta, \frac{Q_i}{P_i}\right) - \frac{\eta^2\zeta_1}{P_i^2}e^{\frac{-\eta Q_i}{P_i}}\frac{\partial\psi}{\partial Q_i} + \frac{\eta\zeta_1 P_i^{\beta+1}}{Q_i^2}e^{\frac{-\eta Q_i}{P_i}}\frac{\partial\psi}{\partial P_i}$$

$$+ \frac{\eta^2\zeta_1 P_i^{\beta}}{Q_i}e^{\frac{-\eta Q_i}{P_i}}\frac{\partial\psi}{\partial P_i} - \frac{\eta\zeta_1 P_i^{\beta+1}}{Q_i}e^{\frac{-\eta Q_i}{P_i}}\frac{\partial^2\psi}{\partial Q_i\partial P_i}$$

$$= \zeta_5 \text{ (say)}$$

and

$$\frac{\partial^2 L}{\partial Q_i \partial\eta} = -\frac{\zeta_1 P_i^{\beta+1}}{Q_i^2}e^{\frac{-\eta Q_i}{P_i}}\psi\left(\eta, \frac{Q_i}{P_i}\right) + \frac{\eta\zeta_1 P_i^{\beta}}{Q_i}e^{\frac{-\eta Q_i}{P_i}}\psi\left(\eta, \frac{Q_i}{P_i}\right) - \frac{\zeta_1 P_i^{\beta+1}}{Q_i}e^{\frac{-\eta Q_i}{P_i}}\frac{\partial\psi}{\partial Q_i}$$

$$- \eta\zeta_1 P_i^{\beta-1}e^{\frac{-\eta Q_i}{P_i}}\psi\left(\eta, \frac{Q_i}{P_i}\right) + \zeta_1 P_i^{\beta}e^{\frac{-\eta Q_i}{P_i}}\frac{\partial\psi}{\partial Q_i} + \frac{\eta\zeta_1 P_i^{\beta+1}}{Q_i^2}e^{\frac{-\eta Q_i}{P_i}}\frac{\partial\psi}{\partial\eta} + \frac{\eta^2\zeta_1 P_i^{\beta}}{Q_i}e^{\frac{-\eta Q_i}{P_i}}\frac{\partial\psi}{\partial\eta}$$

$$- \frac{\eta\zeta_1 P_i^{\beta+1}}{Q_i}e^{\frac{-\eta Q_i}{P_i}}\frac{\partial^2\psi}{\partial Q_i\partial\eta}$$

$$= \zeta_6 \text{ (say)}.$$

Now, taking a derivative of Equation (14) with respect to η, one has

$$\frac{\partial^2 L}{\partial P_i \partial\eta} = -\frac{\zeta_1(\beta+1)P_i^{\beta}}{Q_i}e^{\frac{-\eta Q_i}{P_i}}\psi\left(\eta, \frac{Q_i}{P_i}\right) - \eta\zeta_1 P_i^{\beta-1}e^{\frac{-\eta Q_i}{P_i}}\psi\left(\eta, \frac{Q_i}{P_i}\right) - \frac{\zeta_1 P_i^{\beta+1}}{Q_i}e^{\frac{-\eta Q_i}{P_i}}\frac{\partial\psi}{\partial P_i}$$

$$+ \eta\zeta_1\beta P_i^{\beta-1}e^{\frac{-\eta Q_i}{P_i}}\psi\left(\eta, \frac{Q_i}{P_i}\right) + \eta^2\zeta_1 Q_i P_i^{\beta-2}e^{\frac{-\eta Q_i}{P_i}}\psi\left(\eta, \frac{Q_i}{P_i}\right) + \eta\zeta_1 P_i^{\beta}e^{\frac{-\eta Q_i}{P_i}}\frac{\partial\psi}{\partial P_i}$$

$$- \frac{\eta\zeta_1(\beta+1)P_i^{\beta}}{Q_i}e^{\frac{-\eta Q_i}{P_i}}\frac{\partial\psi}{\partial\eta} - \eta^2\zeta_1 P_i^{\beta-1}Q_i e^{\frac{-\eta Q_i}{P_i}}\frac{\partial\psi}{\partial\eta} - \frac{\eta\zeta_1 P_i^{\beta+1}}{Q_i}e^{\frac{-\eta Q_i}{P_i}}\frac{\partial^2\psi}{\partial P_i\partial\eta}$$

$$= \zeta_7 \text{ (say)}.$$

$\square$

To show the global maximum, the principle minors must be alternative in sign. Thus, the conditions are made like this way: either (i) satisfies or (ii) satisfies; then, the profit function contains a global maximum at the optimum value of the decision variables.

The model is tested through numerical experiments and the global optimality is tested at the optimal points.

5. Numerical Experiment

This section consists of numerical examples and sensitivity of the model.

5.1. Numerical Examples

There are two examples in this section.

5.2. Example 1

The following parametric values are taken from [2] as in Table 2. The model considers for the Example 1 with two items. Figures 3–8 indicate the variation in development cost with failure rate, variation in expected total profit per unit time with other cost key parameters, expected total profit versus lot size of two products, expected total profit versus production rate of two products, expected total profit versus lot size of first product and failure rate, and expected total profit versus production rate of 1st product and failure rate, respectively.

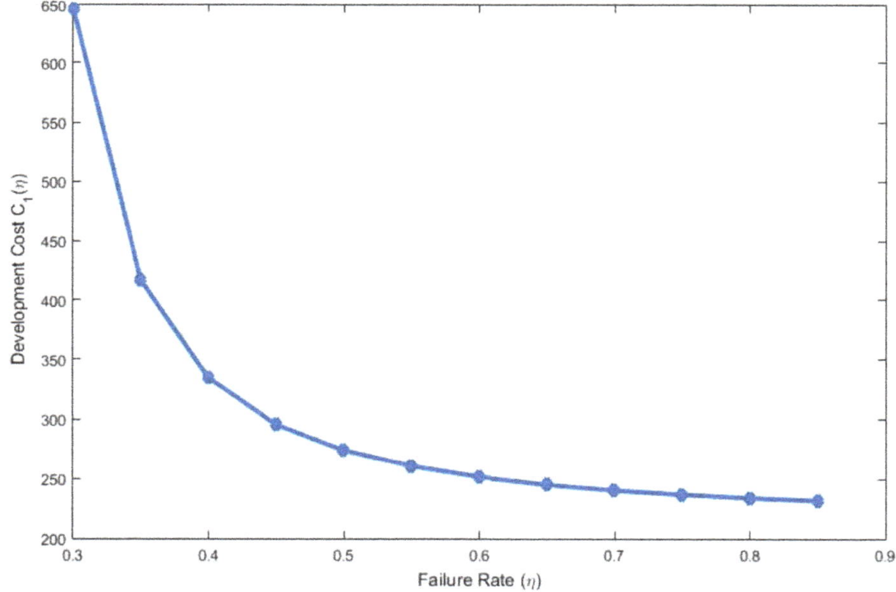

Figure 3. Variation in development cost $C_1(\eta)$ with varying system failure rate (η).

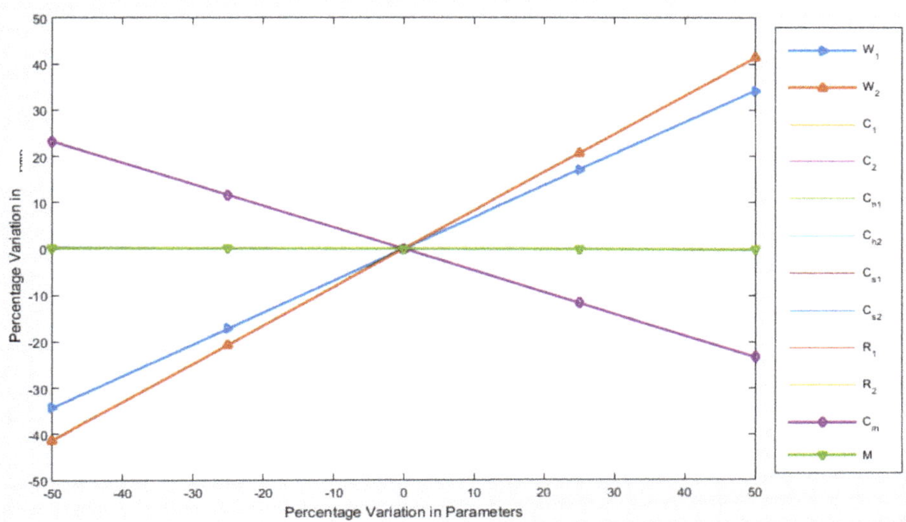

Figure 4. Variation in expected total profit per unit time (ETP) with variation in system parameters (selling price of first product w_1, selling price of second product w_2, inspection cost of first product C_1, inspection cost of second product C_2, holding of first product C_{h_1}, holding of second product C_{h_2}, setup cost of first product C_{s_1}, setup cost of second product C_{s_2}, rework cost of first product R_1, rework cost of second product R_2, material cost C_m, and fixed cost M).

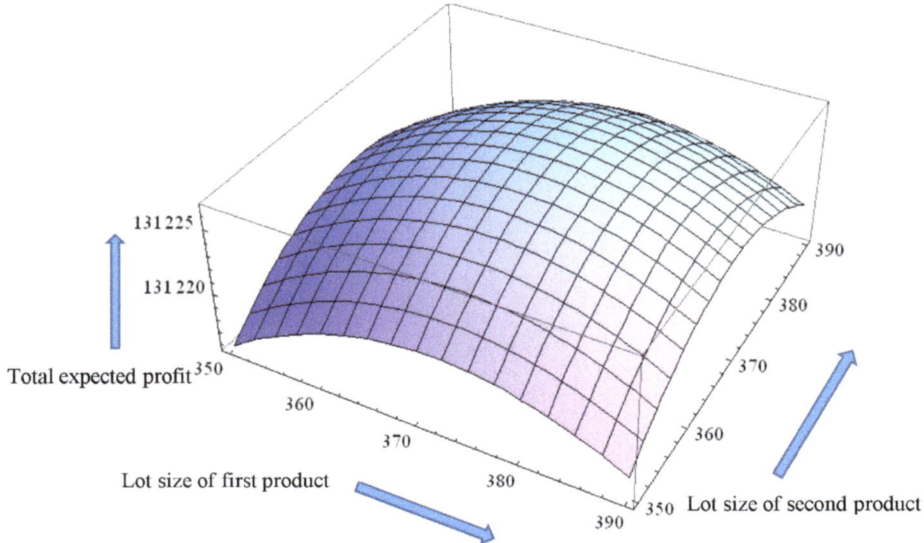

Figure 5. Expected total profit (ETP) versus lot size of two products (Q_1, Q_2).

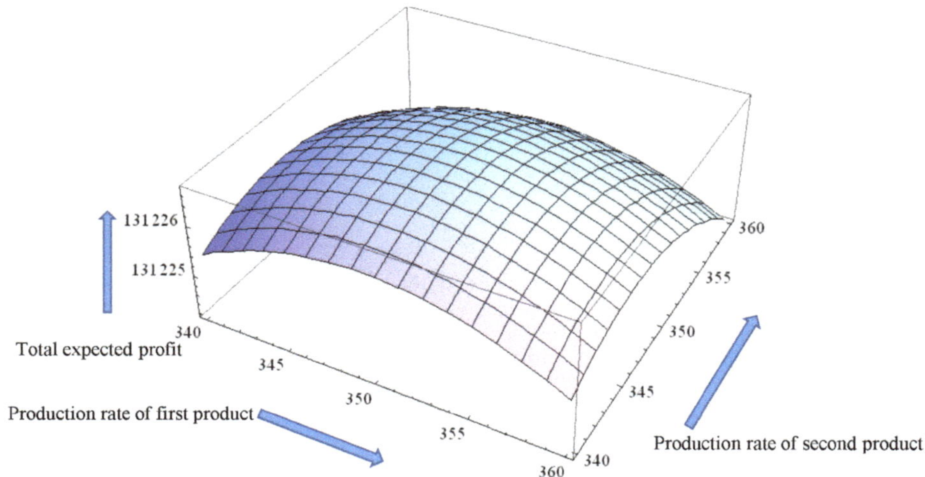

Figure 6. Expected total profit (ETP) versus production rate of two products (P_1, P_2).

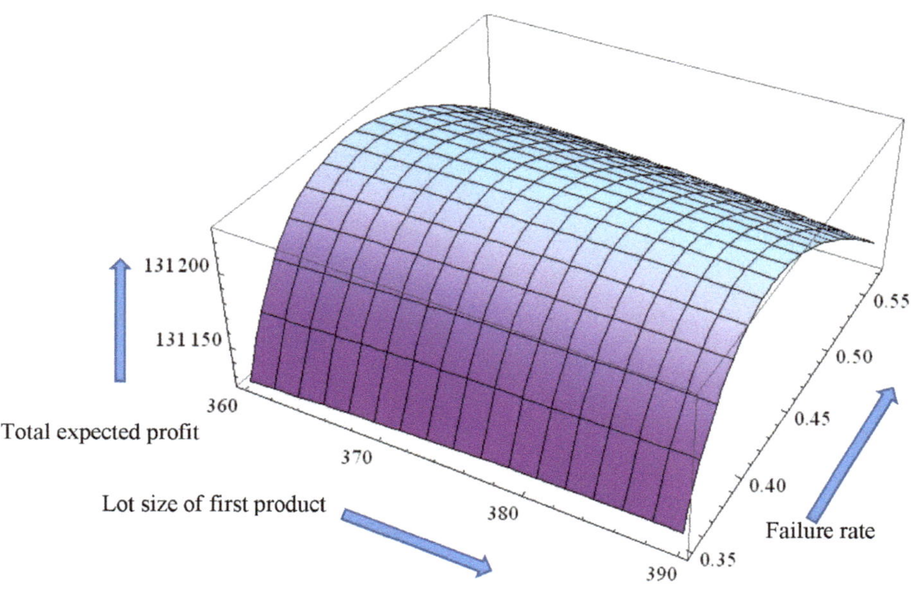

Figure 7. Expected total profit (ETP) versus lot size of first product Q_1 and failure rate η.

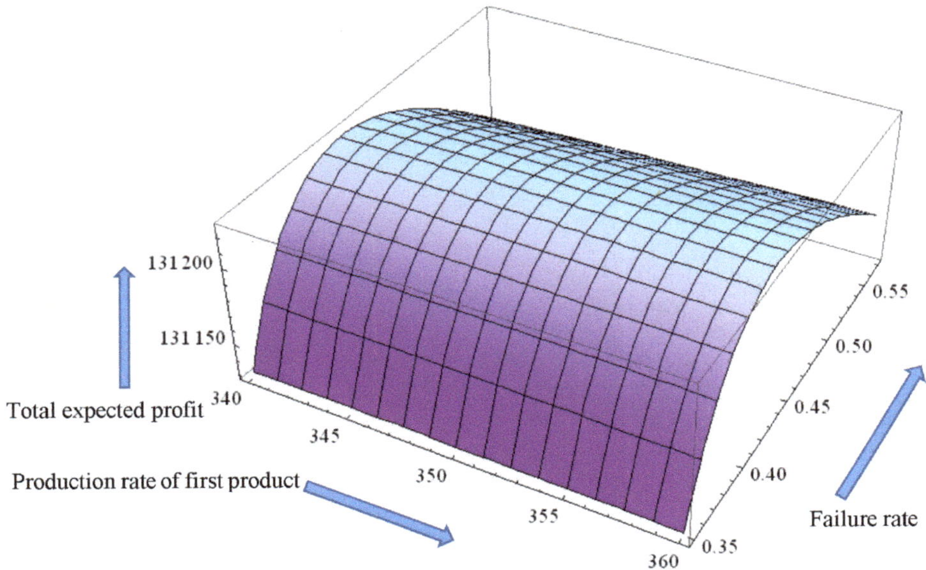

Figure 8. Expected total profit (ETP) versus production rate P_1 of 1st product and failure rate η.

Table 2. The optimal values of Example 1.

i	D_i units	W_i \$/unit	C_i \$/unit	C_{h_i} \$/unit/year	C_{s_i} \$/setup	R_i \$/unit	ξ_i sq. feet/unit	ϕ_i \$/unit
1	300	300	2.7	1.8	920	95	5	10
2	310	350	1.8	2.1	1000	102	4	12

i	C_{s_i}' \$/setup	C_{h_i}' \$/unit/year	C_i' \$/unit	R_i' \$/unit
1	80	0.2	0.3	5
2	100	0.4	0.2	8

M \$/unit	C_m \$/unit	X \$	A sq.feet	B \$	r	α	β	γ	δ
200	100	30	10,000	15,000	0.90	0.05	0.25	2	0.70

The results of Example 1 are given in Table 3.

Table 3. The optimal values of Example 1.

ETP^* \$/year	Q_1^* units	Q_2^* units	P_1^* unit/year	P_2^* unit/year	η^*
131,219	371.64	341.91	349.05	297.99	0.43

5.3. Example 2

For Example 2, the model considers three items. The following parametric values are taken from Sana (2010b) as in Table 4. The model considers for Example 2 with three items.

Table 4. Input parameters for $i = 3$.

i	D_i units	W_i \$/unit	C_i \$/unit	C_{h_i} \$/unit/year	C_{s_i} \$/setup	R_i \$/unit	ξ_i sq. feet/unit	ϕ_i \$/unit
1	300	300	2.7	1.8	920	95	5	10
2	310	350	1.8	2.1	1000	102	4	12
3	330	400	3.6	2.5	1050	120	6	14

i	C_{s_i}' \$/setup	C_{h_i}' \$/unit/year	C_i' \$/unit	R_i' \$/unit
1	80	0.2	0.3	5
2	100	0.4	0.2	8
3	150	0.5	0.4	10

M \$/unit	C_m \$/unit	X \$	A sq.feet	B \$	r	α	β	γ	δ
200	100	30	$10,000$	$15,000$	0.90	0.05	0.25	2	0.70

The results of Example 2 are given in Table 5.

Table 5. The optimal values of Example 2.

ETP^* \$/year	Q_1^* units	Q_2^* units	Q_3^* units	P_1^* unit/year	P_2^* unit/year	P_3^* unit/year	η^*
$226,158$	372.69	372.39	372.19	349.11	347.19	355.16	0.43

5.4. Sensitivity Analysis

The sensitivity analysis of key parameters are considered and major findings can be concluded from the sensitivity analysis Table 6.

Table 6. Sensitivity analysis of key parameters.

Parameters	Changes (in %)	ETP (%)	Parameters	Changes (in %)	ETP (%)
W_1	-50%	-34.29	C_{s1}	-50%	$+0.36$
	-25%	-17.15		-25%	$+0.16$
	$+25\%$	$+17.15$		$+25\%$	-0.15
	$+50\%$	$+34.29$		$+50\%$	-0.28
W_2	-50%	-41.34	C_{s2}	-50%	$+0.40$
	-25%	-20.67		-25%	$+0.19$
	$+25\%$	$+20.67$		$+25\%$	-0.17
	$+50\%$	$+41.34$		$+50\%$	-0.32
C_1	-50%	$+0.34$	R_1	-50%	$+0.14$
	-25%	$+0.17$		-25%	$+0.06$
	$+25\%$	-0.17		$+25\%$	-0.04
	$+50\%$	-0.34		$+50\%$	-0.08
C_2	-50%	$+0.24$	R_2	-50%	$+0.16$
	-25%	$+0.12$		-25%	$+0.07$
	$+25\%$	-0.12		$+25\%$	-0.05
	$+50\%$	-0.24		$+50\%$	-0.09
C_{h1}	-50%	$+0.03$	M	-50%	$+0.14$
	-25%	$+0.01$		-25%	$+0.07$
	$+25\%$	-0.007		$+25\%$	-0.07
	$+50\%$	-0.01		$+50\%$	-0.13
C_{h2}	-50%	$+0.03$	C_m	-50%	$+23.24$
	-25%	$+0.01$		-25%	$+11.62$
	$+25\%$	-0.006		$+25\%$	-11.62
	$+50\%$	-0.006		$+50\%$	-23.24

Table 6 is showing the effect of changes by certain percentage (-50%, -25%, $+25\%$, $+50\%$) of the key parameters and selling-price on the optimal values of total expected profit.

The following are some inferences from obtained results:

1. The higher the selling-price, the higher the optimal value of the profit and vice versa. It is generally true that, keeping all costs the same, and an increase in selling-price increases the profit. However, in specific cases, where demand depends upon the selling-price, there could be an optimal point beyond which, if selling-price is increased, the demand and hence profit could decrease. As demand is not price-dependent in the considered model, profit can thus be increased by increasing product selling-price.

2. On the other hand, the increase in inspection cost decreases the optimal value of the profit. Some industries do not consider inspection as a value added process to the final product, thus the investment in inspection does not give value apparently. Investing more for inspection results in losing profit. Likewise, reworking cost is another such type of cost that is considered non-value added to the final product because the investment is being done again to make the product saleable. Reworking cost also affects the profit in the same way as the inspection cost does.

3. Observing the data in Table 6, it can be concluded generally that variation in holding cost has a less significant effect on the optimal profit of the system. However, if it is the reverse, but there is very little variation in the profit. The results are justified because holding cost is very small as compared to other costs for this model as products are not holding for a long-time. Similarly, variation in setup cost has a nominal effect on optimality of profit.

4. Variation in material cost varies the optimal value of total profit significantly as it is obvious from Table 6. In a production system, the major costs are incurred by material and production. Working on these areas, it can reduce the total cost to a considerable amount and hence it can increase the profit.

5. Effect of fixed cost changes are the reverse for the optimal value of the total expected profit, whereas the effect is very nominal.

6. It can be summarized that the optimal profit value is highly sensitive to the selling-price and material cost, while there are nominal changes in optimal profit for the changes in other parameters. From a managerial point of view, concerned with the total amount of profit, the optimal selling-price and material cost are the most important factors to control its optimal value.

5.5. Managerial Insights

Some recommendations are given for the industry as follows:

1. The major insight for the industry manager is that they have smart production systems without having the information about the benefit of energy consumption calculation. Basically, for a smart production system, a portion of the total amount of labor is replaced by some smart machines as results, the consumption of energy increases. However, no research has considered the multi-item smart production with energy consumption and random failure rate yet. Thus, the result of this study would support the smart industry manager for obtaining more profit.

2. As the study is considered for a smart production system, the random breakdown time is still considered. Thus, the smart industry manager can easily calculate the due date for any delivery based on the available data for the random breakdown time. They might get support if they have any issues for budget and space.

3. The smart managers always obtain the profit at the optimum values of the decision variables. They can obtain the information easily about the cost for the energy sector how much energy can be used for which sector. They can decide on the amount of optimized energy consumption level or not. If they found that the energy cost is getting high, they can consider the replacement of this energy consumption with any other renewable energy. This can benefit for the random defective rate information during long-run smart production systems.

6. Conclusions

This study extended a basic production model with some realistic assumptions explained here.

- The inspection cost for multi-stage smart production systems with smart machines under energy consumption is not negligible.
- The development cost is not constant for any multi-stage smart production system. This study considered that the development cost was dependent on the labor cost, energy resource cost and the failure rate of the smart production system.
- A multi-item smart production system without budget and space constraints is almost impossible. Thus, this model considered a space and a budget constraint to match with reality.
- The energy consideration in multi-item smart production systems had not yet been considered by authors. This study introduced the consideration of energy in a smart production systems under budget and space constraints for the first time.

It was considered that system reliability depends upon system failure rate and the system reliability was considered as a system design variable. The greater the investment, the more reliable the the system would be. Investment for the system reliability was done for production costs that are composed of development costs, material costs and tool/die costs. Production of defective items was dependent upon the state of systems, when it was moved from *in-control* state to *out-of-control* state and the movement time was supposed as random, which followed an exponential distribution, and an expected number of defective items was calculated during the production cycle under energy consideration. The mathematical model for expected total profit was formulated and solved using the Kuhn–Tucker method considering system constraints. To show the practical implications of the model, numerical examples have been solved to compute the optimal value of the objective function and that of decision variables. Finally, sensitivity analysis was presented to study the effect of different system parameters on the optimal value of decision variables and that of objective function. The industry manager obtained the benefit of having information of the energy consumption from all workstations of multi-item smart production systems. They obtained the expected schedule of due dates as they had the information about random defective rates even though smart machines were used. The main limitation of the model is that the demand is known as constant, which may be random or uncertain based on the real-life situation. The smart production system is considered, but autonomation policy is not adopted under the effect of energy. The labor cost is incorporated, but the quality of labor i.e., skilled, semi-skilled, or unskilled is not considered. Those are the main limitations in the direction of energy, which can be considered for further study of this research model. The model can be extended by considering the concept of [15] as imperfect inspection and non-inspected products with warranty. The preventive and corrective maintenance can be another major extension of the model considering planned backorder. This model can be extended further (see, for reference, [1]).

Author Contributions: Data Curation, Software, M.S. and W.I.; Formal Analysis, Investigation, Writing—Original Draft Preparation, M.S.; Conceptualization, Methodology, Validation, Visualization, Funding Acquisition, Supervision, Project Administration, Writing—Review and Editing, B.S.

Funding: This research received no external funding.

Conflicts of Interest: The authors declare no conflict of interest.

Nomenclature

Decision variables

Q_i	lot size of ith product, $i = 1, 2, ..., n$, (units)
P_i	production rate for ith product, $i = 1, 2, ..., n$, (unit/year)
η	failure rate, indicates system reliability, known as system design variable

Random variables

τ_i random time of system moves from *in-control* state to *out-of-control* state for ith item, $i = 1, 2, ..., n$

Parameters

$C_1(\eta)$ development cost per unit item (\$/unit)

C_p production cost per unit item (\$/item)

D_i demand for ith product, $i = 1, 2, ..., n$, (units)

C_{hi} holding cost of the ith product per unit per unit time, $i = 1, 2, ..., n$, (\$/unit/unit time)

C_{si} setup cost of ith product per setup, $i = 1, 2, ..., n$, (\$/setup)

C_i inspection cost per item (\$/item)

T length of production cycle (year)

t_{1i} time required for maximum inventory, $i = 1, 2, ..., n$, (year)

C_m material cost per unit item (\$/unit)

R_i rework cost of ith item, $i = 1, 2, ..., n$, (\$/ defective item)

N number of defective items in a production cycle

M fixed labor and energy costs, independent of reliability η (\$/item)

η_{max} maximum value of system failure rate

η_{min} minimum value of system failure rate

$E(N)$ expected number of defective items per unit production cycle

ETP total expected profit per unit time (\$/year)

W_i selling-price of the ith product, $i = 1, 2, ..., n$, (\$/unit)

X fixed cost for resources (\$)

ϕ_i budget consumed on unit item of product i, $i = 1, 2, ..., n$, (\$/unit)

ξ_i space occupied of a unit item of product i, $i = 1, 2, ..., n$, (square feet/item)

A maximum space available for storing (square feet)

B maximum budget available (\$)

Abbreviations

The following abbreviations are used in this manuscript:

NPV	Net present value
EMQ	Economic manufacturing quantity
JIT	Just-in-time
ETP	Expected total profit
UPC	Unit production cost
DC	Development cost
RC	Rework cost
IC	Inspection cost
HC	Holding cost
SC	Setup cost
NA	Not applicable

Appendix A

$$\psi\left(\eta, \frac{Q_i}{P_i}\right) = \frac{t_1^{\gamma+2}}{\gamma+2} + \frac{\eta t_1^{\gamma+3}}{\gamma+3} + \frac{\eta^2 t_1^{\gamma+4}}{2!(\gamma+4)} + \frac{\eta^3 t_1^{\gamma+5}}{3!(\gamma+5)} + \frac{\eta^4 t_1^{\gamma+6}}{4!(\gamma+6)} + \cdots$$

$$= \sum_{j=1}^{\infty} \frac{t_{1i}^{\gamma+j+1} \eta^{j-1}}{(j-1)!(\gamma+j+1)}$$

$$= \sum_{j=1}^{\infty} \frac{\left(\frac{Q_i}{P_i}\right)^{\gamma+j+1} \eta^{j-1}}{(j-1)!(\gamma+j+1)}.$$

References

1. Cárdenas-Barrón, L.E. Economic production quantity with rework process at a single-stage manufacturing system with planned backorders. *Comput. Ind. Eng.* **2009**, *57*, 1105–1113. [CrossRef]
2. Sana, S.S. A production-inventory model in an imperfect production process. *Eur. J. Oper. Res.* **2010**, *200*, 451–464. [CrossRef]
3. Kim, C.H.; Hong, Y. An optimal production run length in deteriorating production processes. *Int. J. Prod. Econ.* **1999**, *58*, 183–189. [CrossRef]
4. Rosenblatt, M.J.; Lee, H.L. Economic production cycles with imperfect production processes. *IIE Trans.* **1986**, *18*, 48–55. [CrossRef]
5. Giri, B.; Dohi, T. Exact formulation of stochastic EMQ model for an unreliable production system. *J. Oper. Res. Soc.* **2005**, *56*, 563–575. [CrossRef]
6. Sana, S.S.; Goyal, S.K.; Chaudhuri, K. An imperfect production process in a volume flexible inventory model. *Int. J. Prod. Econ.* **2007**, *105*, 548–559. [CrossRef]
7. Chiu, Y.-S.P.; Chen, K.-K.; Cheng, F.-T.; Wu, M.-F. Optimization of the finite production rate model with scrap, rework and stochastic machine breakdown. *Comput. Math. Appl.* **2010**, *59*, 919–932. [CrossRef]
8. González, H.; Calleja, A.; Pereira, O.; Ortega, N.; López de Lacalle, L.N.; Barton, M. Super abrasive machining of integral rotary components using grinding flank tools. *Metals* **2018**, *8*, 24. [CrossRef]
9. Egea, A.J.S.; Deferrari, N.; Abate, G.; Martínez Krahmer, D.; López de Lacalle, L.N. Short-cut method to assess a gross available energy in a medium-load screw friction press. *Metals* **2018**, *8*, 173. [CrossRef]
10. Sarkar, B. An EOQ model with delay in payments and stock dependent demand in the presence of imperfect production. *Appl. Math. Comput.* **2012**, *218*, 8295–8308. [CrossRef]
11. Sarkar, B. An EOQ model with delay in payments and time varying deterioration rate. *Math. Comput. Model.* **2012**, *55*, 367–377. [CrossRef]
12. Sarkar, B. An inventory model with reliability in an imperfect production process. *Appl. Math. Comput.* **2012**, *218*, 4881–4891. [CrossRef]
13. Chakraborty, T.; Giri, B. Lot sizing in a deteriorating production system under inspections, imperfect maintenance and reworks. *Oper. Res.* **2014**, *14*, 29–50. [CrossRef]
14. Sarkar, B.; Cárdenas-Barrón, L.E.; Sarkar, M.; Singgih, M.L. An economic production quantity model with random defective rate, rework process and backorders for a single stage production system. *J. Manuf. Syst.* **2014**, *33*, 423–435. [CrossRef]
15. Sarkar, B.; Saren, S. Product inspection policy for an imperfect production system with inspection errors and warranty cost. *Eur. J. Oper. Res.* **2016**, *248*, 263–271. [CrossRef]
16. Pasandideh, S.H.R.; Niaki, S.T.A.; Nobil, A.H.; Cárdenas-Barrón, L.E. A multi-product single machine economic production quantity model for an imperfect production system under warehouse construction cost. *Int. J. Prod. Econ.* **2015**, *169*, 203–214. [CrossRef]
17. Purohit, A.K.; Shankar, R.; Dey, P.K.; Choudhary, A. Non-stationary stochastic inventory lot-sizing with emission and service level constraints in a carbon cap-and-trade system. *J. Clean. Prod.* **2016**, *113*, 654–661. [CrossRef]
18. Sana, S.S. An economic production lot size model in an imperfect production system. *Eur. J. Oper. Res.* **2010**, *201*, 158–170. [CrossRef]
19. Cárdenas-Barrón, L.E.; Chung, K.-J.; Kazemi, N.; Shekarian, E. Optimal inventory system with two backlog costs in response to a discount offer: Corrections and complements. *Oper. Res.* **2018**, *18*, 97–104. [CrossRef]
20. Tiwari, S.; Jaggi, C.K.; Gupta, M.;Cárdenas-Barrón, L.E. Optimal pricing and lot-sizing policy for supply chain system with deteriorating items under limited storage capacity. *Int. J. Prod. Econ.* **2018**, *200*, 278–290. [CrossRef]
21. Tiwari, S.; Cárdenas-Barrón, L.E.; Goh, M.; Shaikh, A.A. Joint pricing and inventory model for deteriorating items with expiration dates and partial backlogging under two-level partial trade credits in supply chain. *Int. J. Prod. Econ.* **2018**, *200*, 16–36. [CrossRef]
22. Huang, Y.-F.; Lai, C.-S.; Shyu, M.-L. Retailer's EOQ model with limited storage space under partially permissible delay in payments. *Math. Prob. Eng.* **2007**, *2007*, 90873. [CrossRef]
23. Pasandideh, S.H.R.; Niaki, S.T.A. A genetic algorithm approach to optimize a multi-products EPQ model with discrete delivery orders and constrained space. *Appl. Math. Comput.* **2008**, *195*, 506–514. [CrossRef]

24. Hafshejani, K.F.; Valmohammadi, C.; Khakpoor, A. Retracted: Using genetic algorithm approach to solve a multi-product EPQ model with defective items, rework, and constrained space. *J. Ind. Eng. Int.* **2012**, *8*, 1–8. [CrossRef]

25. Mahapatra, G.; Mandal, T.; Samanta, G. An EPQ model with imprecise space constraint based on intuitionists fuzzy optimization technique. *J. Mult.-Valued Log. Soft Comput.* **2012**, *19*, 409–423. [CrossRef]

26. Mohan, S.; Mohan, G.; Chandrasekhar, A. Multi-item, economic order quantity model with permissible delay in payments and a budget constraint. *Eur. J. Ind. Eng.* **2008**, *2*, 446–460. [CrossRef]

27. Hou, K.-L.; Lin, L.-C. Investing in setup reduction in the EOQ model with random yields under a limited capital budget. *J. Inf. Optim. Sci.* **2011**, *32*, 75–83. [CrossRef]

28. Taleizadeh, A.; Jalali-Naini, S.G.; Wee, H.-M.; Kuo, T.-C. An imperfect multi-product production system with rework. *Sci. Iran.* **2013**, *20*, 811–823. [CrossRef]

29. Cárdenas-Barrón, L.E.; Treviño-Garza, G.; Widyadana, G.A.; Wee, H.-M. A constrained multi-products EPQ inventory model with discrete delivery order and lot size. *Appl. Math. Comput.* **2014**, *230*, 359–370. [CrossRef]

30. Du, Y.; Hu, G.; Xiang, S.; Zhang, K.; Liu, H.; Guo, F. Estimation of the diesel particulate filter soot load based on an equivalent circuit model. *Energies* **2018**, *11*, 472. [CrossRef]

31. Todde, G.; Murgia, L.; Caria, M.; Pazzona, A. A comprehensive energy analysis and related carbon footprint of dairy farms, Part 2: Investigation and modeling of indirect energy requirements. *Energies* **2018**, *11*, 463. [CrossRef]

32. Tomić, T.; Schneider, D.R. The role of energy from waste in circular economy and closing the loop concept-Energy analysis approach. *Renew. Sustain. Energy Rev.* **2018**, *98*, 268–287. [CrossRef]

33. Haraldsson, J.; Johansson, M.T. Review of measures for improved energy efficiency in production-related processes in the aluminium industry-from electrolysis to recycling. *Renew. Sustain. Energy Rev.* **2018**, *93*, 525–548. [CrossRef]

34. Xua, L.; Cheng, J.-H.; Liu, P.; Wang, Q.; Xu, Z.-X.; Liu, Q.; Shen, J.-Y.; Wang, L.-J. Production of bio-fuel oil from pyrolysis of plant acidified oil. *Renew. Energy* **2019**, *130*, 910–919. [CrossRef]

Article

Effects of Carbon-Emission and Setup Cost Reduction in a Sustainable Electrical Energy Supply Chain Inventory System

Umakanta Mishra [1], Jei-Zheng Wu [1,*] and Anthony Shun Fung Chiu [2]

[1] Department of Business Administration, Soochow University, 56 Section 1, Kuei-yang Street, Taipei 10048, Taiwan; umakanta.math@gmail.com
[2] Don Antonio Cojuangco Professorial Chair of Industrial Technology University Fellow, Professor and Research Fellow Gokongwei College of Engineering, De La Salle University Manila, Manila 0922, Philippines; anthony.chiu@dlsu.edu.ph
* Correspondence: jzwu@scu.edu.tw

Received: 15 February 2019; Accepted: 28 March 2019; Published: 29 March 2019

Abstract: This article develops a sustainable electricity supply chain mathematical model that assumes linear price-dependent customer demands where the price is a decision variable under setup cost and carbon emission. The sustainable electrical supply chain system contained: (a) power generation; (b) transmission substations; (c) distribution substations; and (d) customer. The production rates depend on the demand rate, and demand for electricity by the customers is dependent on the price of electricity where the electrical energy was generated and transmitted through multiple substations to customers. Moreover, we considered that the capacities of transmission rates, power generation, and distances in between two stations are associated with the distribution costs and transmission cost. Here, we used the theory of inventory to develop a new model and suggested a procedure to deduce an optimal solution for this model. Finally, a numerical example and sensitivity analysis are employed to illustrate the present study and with managerial insights.

Keywords: sustainable electrical energy supply chain; inventory; price-dependent demand; transmission and distribution costs; carbon-emission

---

1. Introduction

1.1. Background of the Research

Generally, a supply chain inventory model defines the retailer-customer relationship. Setup cost plays a crucial role in today's supply chain inventory management system for shipment of items on time. The setup procedure is not evaluated as a fixed/known constraint, but needs to be considered at a time of minimizing waste, satisfying deadlines, productivity, and elaborating resource utilization. To minimize the total cost investment, the manufacturer needs to reduce setup costs. In a supply chain inventory system, the capacity of electrical energy has the same process for determining the order quantity and total profit. In an electrical supply chain inventory system, electricity generated in a power generation plant will be transmitted through a transmission line and distribution substation to the customers whose demand is influenced by the price of electricity in order to maximize the profit.

The consumption of electricity energy has rapidly increased. The total electricity consumption worldwide in 2015 was greater than in 1980. Electricity consumption for industrial, commercial, residential and transportation sectors has been enhanced since 2015. This is due to increasing climate changes upon home electricity use at a rate of more than 100 million kWh/day, due to the need for air-conditioning (mainly in summer time) in homes. The electricity price may have influenced the demand for electricity. The price of electricity is set to reduce the demand. Generally, the price is usually the maximum for commercial customers and the residential sector because of the cost to distribute electricity supply to them. Furthermore, an increase of global greenhouse gas carbon emissions (GGCE) in all countries of world is farther than ever from reaching the goals of the Paris Agreement, according to the new United Nations report, released in 2018, prior to a meeting of officials and environmental experts from all around of the world in Katowice, Poland, for climate negotiations. A few years ago at the environmental negotiations in Paris all countries agreed to bring down GGCE sufficient to maintain global warming below 2 degrees Celsius, and under 1.5 degrees. A report from the UN's top climate panel published earlier this decrease found that an additional 0.5 degrees of warming would have forceful and unfortunate effects on the environment, allowing for more empirical support for those countries pressing for more challenging targets during the negotiations at COP24. But to maintain temperature increases under 1.5 degrees, global carbon emissions are required to peak by 2020. In between 2014 to 2016, global carbon emissions stayed comparatively flat, and negotiators hope that the decade long trend of increasing carbon emissions is about to reverse. But after few years of stagnation, global carbon emissions grew again in 2017, reaching a record 53.4 gigatons of carbon emission equivalent. To meet the 1.5-degree mark, the world requires to jointly bring emissions down in the next 12 years. Therefore, governments all over the world should implement cap-and-trade regulations policies to keep the emissions low. For reducing carbon emissions, production companies can monitor and enhance the emission performance of their products during their life-cycle stages. The carbon-emission assessment provides a possible mechanism to serve companies with some emission reduction. During the production process, the manufacturer should formulate a low carbon system.

1.2. Research Questions, Motivation and Contribution of the Model

In the literature review section (Section 1.3) it can be clearly seen that some works seek to determine sustainable electrical supply chain inventory models and no research tries to determine a sustainable electrical supply chain inventory model with setup cost reduction and CO2 emissions. Our research questions this model: (1) what is the electrical power distribution factor's effect on the distribution substation, and the electrical power transmission factor's impact on the transmission substation and electrical power generation factor; (2) how much is the ordering quantity?; (3) what is customer's average electricity consumption time?; and (3) what is the retailer's selling price? To answer these questions in this study, we developed and solved a sustainable electricity supply chain inventory model with setup cost reduction and CO2 emissions while considering environmental parameters. The aim of this study is to examine the effect of price-dependent demand on the sustainable electrical supply chain inventory system under setup cost reduction and carbon emissions. The energy is transmitted via distribution networks to customers whose demand is determined by the price of electricity in order to find an optimal solution. The contributions of this paper are presented in Table 1.

1.3. Literature Review

Several research papers highlighted various integrated inventory models with various key parameters. Yang [1] deduced an inventory model considering lead time and crashing cost. Hoque [2] studied an integrated inventory model considering a normal distribution lead time, setup time, batch time and the cost of transportation. Sarkar et al. [3] deduced an inventory model by adding an imperfect production concept. A non-defective product adopts a binomial distribution function and demand adopts a mixture of the normal distribution function in their model. Mishra [4] formulated a production-inventory model with price dependent demand where the production depends on the rate of demand. Multiple buyers and a single-vendor model apply in a food inventory system studied by Fauza et al. [5]. Denizel et al. [6] formulated a lot size inventory model with setup costs decreased by various amounts depending upon the raw-materials and discussed the shortest path problem. Diaby [7] demonstrated a complete model to reduce setup cost and time with determining cut setup time and minimizing the total cost. Nyea et al. [8] studied several inventory models for optimal investment of setup cost reduction or optimal setup times; and also discussed the queuing model to estimate work in the process level in their models. Later, Freimer et al. [9] demonstrated improvement in quality and setup cost reduction of the process in their model. Huang et al. [10] assumed setup cost policy reduction by an added investment cost. Sarkar and Moon [11] deduced a model by putting the policy of reorder point, quality improvement and lead time by considering that backorder rate has a great impact under a production process and is imperfect. Sarkar et al. [12] studied the concept of setup cost-reduction policy under quality improvement. Sarkar et al. [13] studied an effect of setup cost-reduction process in a two-echelon supply chain inventory model under deterioration and using the technique of quality improvement. Sarkar et al. [14] developed an integrated inventory model for a setup cost-reduction policy, carbon-emission policy and used the technique of the Stackelberg game approach to find the total cost. Little research has been done on a supply chain electrical energy inventory model. Banbury [15] was the first researcher to developed an electricity supply chain model. Thereafter, (Schneider et al. [16], Schneider et al. [17]) first developed this using a very simple inventory policy to find the electrical supply chain policy. Later, Taylor et al. [18] studied capacity and price competition in an electricity market by using a two-stage game theory model. Wu et al. [19] studied a new model to examine the formal generation with sporadic supply. Ouedraogo [20] formulated an electricity supply chain with demand in an African power system. Wangsa and Wee [21] studied an electrical supply chain inventory policy assuming the blackout cost. Recently, interesting research by Wangsa et al. [22] assumed a sustainable supply chain inventory model and the effect of price-dependent demand.

Table 1. Other authors contribution to this theme.

Author	Inventory Model	Electrical Supply Chain System	Setup Cost Reduction	Carbon Emission
Yang [1]	✓	×	×	×
Hoque [2]	✓	×	×	×
Sarkar et al. [3]	✓	×	×	×
Fauza et al. [5]	✓	×	×	×
Denizel et al. [6]	×	×	✓	×
Diaby [7]	×	×	✓	×
Nyea et al. [8]	×	×	✓	×
Freimer et al. [9]	×	×	✓	×
Huang et al. [10]	✓	×	✓	×
Sarkar and Moon [11]	×	×	✓	×
Sarkar et al. [12]	×	×	✓	×
Sarkar et al. [14]	✓	×	✓	✓
Banbury [15]	×	✓	×	×
Schneider et al. [16]	✓	✓	×	×
Schneider et al. [17]	✓	✓	×	×
Taylor et al. [18]	×	✓	×	×
Wu et al. [19]	×	✓	×	×
Ouedraogo [20]	×	✓	×	×
Wangsa and Wee [21]	✓	✓	×	×
Wangsa et al. [22]	✓	✓	×	×
Hammami et al. [23]	✓	×	×	✓
Tang et al. [24]	×	×	×	✓
Tang et al. [25]	×	×	×	✓
Ouyang et al. [26]	✓	×	×	×
This paper	✓	✓	✓	✓

Finally, in this study we investigate many research articles involving supply chain inventory model-related carbon emissions, setup cost reduction and electricity energy. Research paper related to the above are the following: Hammami et al. [23] developed a multi-echelon supply chain model with reducing carbon emission, several manufacturing facilities, different outside suppliers, and distinct distribution centers. Tang et al. [24] studied a carbon-emission policy with minimal frequency of shipments for a periodic inventory review system. Thereafter, Tang et al. [25] developed a sustainable supply chain network for consumers, and environmental manners are added by inventory, routing, and location.

1.4. Methodology

In this section, we consider a sustainable electricity supply chain power system with price-dependent demand electricity demand. In reality, it is shown that determining the capacity of a sustainable electrical supply chain system has the same methodology as finding the order quantity q in the supply chain inventory system. The supply chain inventory system involves a vendor–buyer coordination and freight forwarding. The buyer sells items to the customers and orders items from the vendor. The vendor produces the items and sends in batch to the buyer. The buyer will then sell the items to the customers. But in sustainable electrical supply chain system case, the electricity generated from a power generation will be transmitted through a transmission line and distribution substation to the customers whose demands are influenced by the price of electricity where the electricity is continuously supplied to consumers without any interruptions. The electricity demand is considered as $D(p)$ kWh/year. The power generation produces the electricity in a batch size of $qT\zeta\eta\rho$ kWh where ζ (positive integer) is the distribution factor's effect on the distribution substation, η (positive integer) is the transmission factor's impact on the transmission substation and ρ (positive integer) is a power generation factor. The finite power supply rate is $P = \lambda D(p)$ kWh/year, $[\lambda > 1]$ and a setup cost. The electricity energy of $qT\zeta\eta$ kWh is supplied by the power generator to the transmission substation, then $qT\zeta$ kWh of electricity is supplied to the distribution substation and $E = qT$ kWh of electricity is consumed by the customers. Hence, to maximize the profit of a sustainable electrical supply chain system, we consider the sales revenue, production cost, setup cost reduction of the power generation, ordering cost of customers and transmission/distribution costs of substations. The transmission and distribution costs are functions of the power plant, the transmission substation and the distribution substation with maximum capacities of in z_p^x kWA, z_t^x kWA, z_d^x kWA, respectively. The comparison in supply chain inventory system and sustainable electrical supply chain system are shown in Figure 1.

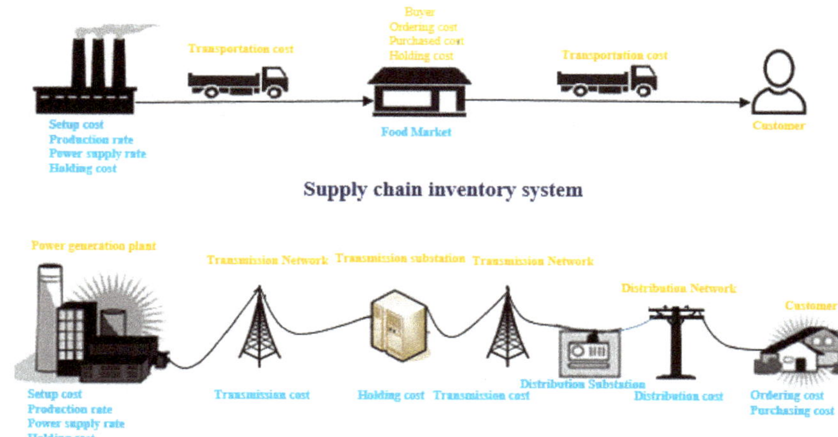

Figure 1. Supply chain inventory and sustainable electrical supply chain.

The rest of the paper is prepared as follows: in Section 2, the assumption and notation for model formulation of the electrical supply chain inventory system are given; in Section 3, we describe the model formulation of the electrical supply chain inventory system. In Section 4, a numerical example is presented for validation of this model. A discussion of the managerial implications is presented in Section 5. Section 6 concludes the work by offering directions for future research work.

2. Assumptions and Notations

In this section, some notation can be used for development of the mathematical model (see Section 3) using the following assumptions:

2.1. Assumptions

- An integrated inventory model is considered.
- A single-buyer for single-types of items are considered.
- Power supply blackouts are not considered.
- The finite power supply rate is greater than demand rate and power supply and demand relation is $P = \lambda D(p)$, where $\lambda > 1$.
- Demand is a linear function of the selling price, which is a more realistic representation of the real world than is the assumption that the demand is a fixed parameter. The linear demand function is considered to be $D(p) = a - bp$ (as shown in Mishra et al. [27]; Mishra et al. [28]) and the electricity demand rate of the customers depends on selling price; where $a > 0$ is scaling factors, $b > 1$ is the elasticity coefficient and satisfied the condition $p < \frac{a}{b}$.
- To reduce setup cost, investment cost I is considered. The expression of setup cost $S(I) = v_0 e^{-\tau I}$, where $v_0 (> 0)$ setup cost at the initial stage and $\tau (> 0)$ is a constant parameter. $dS(I)/dI = -\tau v_0 e^{-\tau I}, d^2 S(I)/dI^2 = \tau^2 v_0 e^{-\tau I} > 0$, means that if the investment will be a higher value than the setup cost it will be smaller value. Therefore, the investment function, setup cost for every production run can be lower value. The investment I decisions will consequence on the setup cost and the setup cost can be a major function of the production system. For example, see Sarkar et al. (2016).
- The total power consumption is q in kW.
- $E = qT$ kWh of electricity is consumed by the customers within a particular time T.
- $qT\zeta$ kWh of electricity is supplied to the distribution substation; where ζ (positive integer) is the distribution factor's effect on the distribution substation.
- $qT\zeta\eta$ kWh is the electricity energy supplied by the power generator to the transmission substation; where ζ (positive integer) is the distribution factor's effect on the distribution substation and η (positive integer) is the transmission factor's impact on the transmission substation.
- $qT\zeta\eta\rho$ kWh of power generation produces the electricity in a batch size; where ζ (positive integer) is the distribution factor's effect on the distribution substation, η (positive integer) is the transmission factor's impact on the transmission substation and ρ (positive integer) is a power generation's factor.
- Power generation, transmission and distribution are following functions; Maximum capacity power generation plant is z_p^x in kWA. Maximum capacity the transmission substation is z_t^x in kWA. Maximum capacity the distribution substation is z_d^x in kWA.
- The relations of the maximum capacity of power generation, transmission substation and distribution substation is $z_p^x > z_t^x > z_d^x$.
- The total power generation process should include the transmission and distribution costs.

2.2. Notation

Notations are used in the model are shown in Table 2 as follows:

Table 2. Notations.

Parameter/Decision Variable/Function	Notations	Descriptions	Units
Parameter	A	Ordering cost per order.	$/unit/order
	a	Scaling factor.	units
	b	Price elasticity coefficient.	units
	k	Production cost.	$/order/year
	v_0	Setup cost at the initial stage.	$/setup
	τ	Known parameter of setup cost $S(I)$.	unit
	r_1	Annual percent of energy holding cost of the transmission substation.	$/unit/year
	r_2	Annual percent of energy holding cost of power generation.	$/unit/year
	Ω	Power factor correction.	kVA/kWh
	c_1	The per mile transmission and distribution rates.	$/kVA/mile
	m_1	The per mile transportation network from power generation to the transmission substation.	miles
	m_2	The per mile transportation network from the transmission substation to the distribution substation.	miles
	m_3	The per mile transportation network from the distribution substation to the customers.	miles
	z_p^x	Maximum capacity of power generation.	kVA
	z_t^x	Maximum capacity of transmission substation.	kVA
	z_d^x	Maximum capacity of distribution substation.	kVA
	β	The power supply loss factor ($0 \leq \beta \leq 1$).	units
Emission parameter	$\hat{r}_1$	Per unit carbon emission of energy holding of the transmission substation.	ton/year
	$\hat{r}_2$	Per unit carbon emission of energy holding of power generation.	ton/year
	$\hat{c}_1$	Per unit carbon emission of transmission and carbon emissions distribution rates.	ton/kVA/mile
	ξ	Carbon tax per unit (money units for each unit of carbon emitted as tax)	$/ton/year
Decision variable	I	Investment cost for setup of power generation.	$/setup
	ζ	The electrical power distribution factor's effect on the distribution substation (positive integer).	units
	η	The electrical power transmission factor's impact on the transmission substation (positive integer).	units
	ρ	The electrical power generation's factor (positive integer).	units
	T	Customer's average electricity (positive integer).	Consumption in hour.
	p	Selling price of electricity (positive integer).	$/kWh.
	q	The customer's power	kW
Function	Π	Total profit.	$/year
	$D(p)$	Demand of customer.	kWh/year
	z_d^y	Actual capacity of the distribution substation.	KVA
	z_t^y	Actual capacity of the transmission substation.	KVA
	z_p^y	Actual capacity of the power generation.	KVA

3. Model Formulation

In this section, we explain how to find the total cost function with regard to customers, the distribution and transmission substations, carbon emission cost, and the total profit function for power generation. The total cost functions are given by the following components:

3.1. Ordering Cost

Ordering cost is defined as the customer's total cost per unit time:

$$TC_O = \frac{AD(p)}{qT} \tag{1}$$

3.2. Distribution Cost

The same methodology as (Wangsa et al. [22]) to determine the distribution cost is obtained from the distribution substation. The cost of distribution for partial load G can be written as $G = \frac{c_1 z_d^x}{z_d^y}$. The transmission function, increase in rate per kVA/mile when z_d^y increases. The distribution cost per kVA/mile is $G_Z = \beta G + (1-\beta)c_1$, where $0 \le \beta \le 1$ is the coefficient of the adjusted inverse function. Therefore $G_Z = \beta c_1 \left[\frac{z_d^x - z_d^y}{z_d^y} \right] + c_1$. The estimated total cost for the distribution substation as a function of demand, corrections factor and distance yields: $G_D = \left[\beta c_1 \left[\frac{z_d^x - z_d^y}{z_d^y} \right] + c_1 \right] D(p) m_3 \Omega$. The actual power supply capacity is given by $z_d^y = qT\zeta\,\Omega$; therefore, the cost for distribution can be written as in the following equation:

$$TC_D = \frac{D(p)\beta c_1 z_d^x m_3}{qT\zeta} + D(p) m_3 \Omega (1-\beta) c_1 \tag{2}$$

3.3. Carbon Emission Distribution Cost

The carbon emission distribution cost has one component. This carbon emission component is related to total distribution rates. The component is summated and briefly represented by:

$$CE_D = \hat{c_1} \left[\frac{D(p)\xi \beta z_d^x m_3}{qT\zeta} + D(p)\xi m_3 \Omega (1-\beta) \right] \tag{3}$$

3.4. Transmission Substation Cost

This total cost comprises of the energy holding cost and transmission cost. The cost of energy holding at the transmission substation is presented by $\frac{r_1 p \zeta \eta E}{2} = \frac{r_1 p \zeta \eta q T}{2}$. The transmission cost can be calculated as: $G_t = \frac{D(p)\beta c_1 z_t^x m_2}{qT\zeta\eta} + D(p) m_2 \Omega (1-\beta) c_1$. Therefore, the total cost obtained at the transmission substation is the sum of energy holding cost and transmission cost. Therefore, the total transmission substations cost;

$$TC_t = \frac{r_1 p \zeta \eta q T}{2} + \frac{D(p)\beta c_1 z_t^x m_2}{qT\zeta\eta} + D(p) m_2 \Omega (1-\beta) c_1 \tag{4}$$

3.5. Carbon Emission Transmission Substation Cost

The carbon emission transmission substation cost has three components. The first component is related to total energy holding rates at the transmission substation, the second and third components are related to the total transmission rates. These components are summated and briefly represented by

$$CE_t = \frac{\hat{r_1}\xi p \zeta \eta q T}{2} + \frac{D(p)\beta\xi\hat{c_1} z_t^x m_2}{qT\zeta\eta} + D(p) m_2 \Omega (1-\beta)\xi\hat{c_1} \tag{5}$$

3.6. Profit of Power Generation

The total profit of power generation can be presented by:

$$\Pi_{PG} = SR - PC - SC - INVC - HC - TRC - CE_{TR} - CE_{PG} \tag{6}$$

Sales revenue

$$SR = D(p)p \tag{7}$$

Production cost

$$PC = D(p)k \tag{8}$$

Power generation creates electrical energy in $(qT\zeta\eta\rho)$ kWh in one production run. Therefore, the setup cost for power generation can be found by using the equation below. Setup cost is:

$$SC = \frac{D(p)v_0 e^{-\tau I}}{qT\zeta\eta\rho} \tag{9}$$

Total setup investment cost is:

$$INVC = \frac{D(p)I}{qT\zeta\eta\rho} \tag{10}$$

Energy holding cost is:

$$HC = r_2 k \frac{\left[E\zeta\eta\rho\left(\frac{E\zeta\eta}{P} + (\rho - 1) - \frac{E}{D(p)}\right) - \frac{\rho^2 [E\zeta\eta]^2}{2P} \right] - \left[\frac{E\zeta\eta}{D(p)}(1 + 2 + 3\ldots\ldots + (\rho - 1))E \right]}{\frac{E\zeta\eta\rho}{D(p)}}$$

HC can be written as rewritten as $HC = r_2 k \frac{E\zeta\eta}{2}\left[\rho\left(1 - \frac{D(p)}{P}\right) - 1 + \frac{2D(p)}{P}\right]$.
Putting $P = \lambda D(p)$ in HC. Therefore, the holding cost for is

$$HC = r_2 k \frac{qT\zeta\eta}{2\lambda}[\rho(\lambda - 1) - \lambda + 2] \tag{11}$$

Transmission cost can be calculated as:

$$G_p = \frac{D(p)\beta c_1 z_p^x m_1}{qT\zeta\eta} + D(p)m_1\Omega(1 - \beta)c_1 \tag{12}$$

3.7. Total Carbon Emission of Transmission Cost

The carbon emission transmission cost has one component. This transmission emission component is related to total transmission rates. The component is summated and briefly represented by:

$$CE_{TR} = \hat{c}_1\left[\frac{D(p)\xi\beta z_p^x m_1}{qT\zeta\eta} + D(p)\xi m_1\Omega(1 - \beta)\right] \tag{13}$$

3.8. Total Carbon Emission of Energy Holding Cost for Power Generation

The carbon emission energy holding cost has one component. The holding emission component is related to total energy hold in a power generation plant. The component is summated and briefly represented by:

$$CE_{PG} = \hat{r}_2\left[k\frac{qT\xi\zeta\eta}{2\lambda}[\rho(\lambda - 1) - \lambda + 2]\right] \tag{14}$$

Therefore, the total profit for power generation Π_{PG} can be defined as:

$$\Pi_{PG} = D(p)(p - k) - \frac{D(p)v_0 e^{-\tau I}}{qT\zeta\eta\rho} - \frac{D(p)I}{qT\zeta\eta\rho} - (r_2 + \hat{r}_2\xi)k\frac{qT\zeta\eta}{2\lambda}[\rho(\lambda - 1) - \lambda + 2]$$
$$- \left[\frac{D(p)\beta(c_1 + \hat{c}_1\xi)z_p^x m_1}{qT\zeta\eta} + D(p)m_1\Omega(1 - \beta)(c_1 + \hat{c}_1\xi)\right] \tag{15}$$

Therefore, the total profit is:

$$
\begin{aligned}
\Pi &= \Pi_{PG} - TC_O - TC_D - CE_D - TC_t - CE_t \\
&= D(p)\left[p - k - (m_1 + m_2 + m_3)\Omega(1-\beta)(c_1 + \hat{c}_1\xi)\right] \\
&\quad -\frac{D(p)}{qT\zeta\eta\rho}\left[A\zeta\eta\rho + \beta(c_1 + \hat{c}_1\xi)z_d^x m_3\eta\rho + \beta(c_1 + \hat{c}_1\xi)z_t^x m_2\rho + v_0 e^{-\tau I} + I + \beta(c_1 + \hat{c}_1\xi)z_p^x m_1\right] \\
&\quad -\frac{qT\zeta\eta}{2\lambda}\left[\lambda(r_1 + \hat{r}_1\xi)p + (r_2 + \hat{r}_2\xi)k(\rho(\lambda-1) - \lambda + 2)\right]
\end{aligned}
\tag{16}
$$

Next, we analyze the consequence of I, ζ, η, ρ, T and p on Π fixed q by using the second order partial derivatives of Equation (16) with respect to I, ζ, η, ρ, T and p:

$$
\begin{aligned}
\frac{\partial \Pi}{\partial \zeta} &= \frac{D(p)}{qT\zeta^2\eta\rho}\left[\beta(c_1 + \hat{c}_1\xi)z_d^x m_3\eta\rho + \beta(c_1 + \hat{c}_1\xi)z_t^x m_2\rho + v_0 e^{-\tau I} + I + \beta(c_1 + \hat{c}_1\xi)z_p^x m_1\right] \\
&\quad -\frac{qT\eta}{2\lambda}\left[\lambda(r_1 + \hat{r}_1\xi)p + (r_2 + \hat{r}_2\xi)k(\rho(\lambda-1) - \lambda + 2)\right]
\end{aligned}
\tag{17}
$$

$$
\frac{\partial^2 \Pi}{\partial \zeta^2} = -\frac{2D(p)}{qT\zeta^3\eta\rho}\left[\beta(c_1 + \hat{c}_1\xi)z_d^x m_3\eta\rho + \beta(c_1 + \hat{c}_1\xi)z_t^x m_2\rho + v_0 e^{-\tau I} + I + \beta(c_1 + \hat{c}_1\xi)z_p^x m_1\right] < 0
\tag{18}
$$

$$
\begin{aligned}
\frac{\partial \Pi}{\partial \eta} &= \frac{D(p)}{qT\zeta\eta^2\rho}\left[\beta(c_1 + \hat{c}_1\xi)z_t^x m_2\rho + v_0 e^{-\tau I} + I + \beta(c_1 + \hat{c}_1\xi)z_p^x m_1\right] \\
&\quad -\frac{qT\zeta}{2\lambda}\left[\lambda(r_1 + \hat{r}_1\xi)p + (r_2 + \hat{r}_2\xi)k(\rho(\lambda-1) - \lambda + 2)\right]
\end{aligned}
\tag{19}
$$

$$
\frac{\partial^2 \Pi}{\partial \eta^2} = -\frac{2D(p)}{qT\zeta\eta^3\rho}\left[\beta(c_1 + \hat{c}_1\xi)z_t^x m_2\rho + v_0 e^{-\tau I} + I + \beta(c_1 + \hat{c}_1\xi)z_p^x m_1\right] < 0
\tag{20}
$$

$$
\frac{\partial \Pi}{\partial \rho} = \frac{D(p)}{qT\zeta\eta\rho^2}\left[v_0 e^{-\tau I} + I + \beta(c_1 + \hat{c}_1\xi)z_p^x m_1\right] - \frac{qT\zeta\eta(r_2 + \hat{r}_2\xi)k(\lambda-1)}{2\lambda}
\tag{21}
$$

$$
\frac{\partial^2 \Pi}{\partial \rho^2} = -\frac{2D(p)}{qT\zeta\eta\rho^3}\left[v_0 e^{-\tau I} + I + \beta(c_1 + \hat{c}_1\xi)z_p^x m_1\right] < 0
\tag{22}
$$

$$
\begin{aligned}
\frac{\partial \Pi}{\partial T} &= \frac{D(p)}{qT^2\zeta\eta\rho}\left[A\zeta\eta\rho + \beta(c_1 + \hat{c}_1\xi)z_d^x m_3\eta\rho + \beta(c_1 + \hat{c}_1\xi)z_t^x m_2\rho + v_0 e^{-\tau I} + I + \beta(c_1 + \hat{c}_1\xi)z_p^x m_1\right] \\
&\quad -\frac{q\zeta\eta}{2\lambda}\left[\lambda(r_1 + \hat{r}_1\xi)p + (r_2 + \hat{r}_2\xi)k(\rho(\lambda-1) - \lambda + 2)\right]
\end{aligned}
\tag{23}
$$

$$
\frac{\partial^2 \Pi}{\partial T^2} = -\frac{2D(p)}{qT^3\zeta\eta\rho}\left[\begin{array}{l} A\zeta\eta\rho + \beta(c_1 + \hat{c}_1\xi)z_d^x m_3\eta\rho + \beta(c_1 + \hat{c}_1\xi) \\ \times z_t^x m_2\rho + v_0 e^{-\tau I} + I + \beta(c_1 + \hat{c}_1\xi)z_p^x m_1 \end{array}\right] < 0
\tag{24}
$$

$$
\begin{aligned}
\frac{\partial \Pi}{\partial p} &= a - bp - b\left[p - k + (\beta - 1)\Omega(c_1 + \hat{c}_1\xi)(m_1 + m_2 + m_3)\right] - \frac{qT\zeta\eta(r_1 + \hat{r}_1\xi)}{2} \\
&\quad +\frac{b}{qT\zeta\eta\rho}\left[A\zeta\eta\rho + \beta(c_1 + \hat{c}_1\xi)z_d^x m_3\eta\rho + \beta(c_1 + \hat{c}_1\xi)z_t^x m_2\rho + v_0 e^{-\tau I} + I + \beta(c_1 + \hat{c}_1\xi)z_p^x m_1\right]
\end{aligned}
\tag{25}
$$

and

$$
\frac{\partial \Pi}{\partial p} = -2b < 0
\tag{26}
$$

From Equations (18), (20), (22), (24) and (26) it can be concluded that, for any feasible solution of q the total profit function Π (Equation (16) is a concave function of ζ, η, ρ, T and p.

Theorem 1. *For the any positive integer $(\zeta, \eta, \rho, T, p)$, the required objective function Π (Equation (16) is concave function of I and q. Therefore, the maximum value Π (Equation (16) is settled at the point I and q which satisfies $\frac{\partial \Pi}{\partial I} = 0$ and $\frac{\partial \Pi}{\partial q} = 0$.*

Proof. To calculate the optimal solution for fixed-integers $(\zeta, \eta, \rho, T, p)$, used the partial derivatives with respect to I and q, as shown in the following equations:

$$\frac{\partial \Pi}{\partial I} = -\frac{D(p)\left(1 - \tau v_0 e^{-\tau I}\right)}{qT\zeta\eta\rho} \tag{27}$$

$$\frac{\partial \Pi}{\partial q} = \frac{D(Ap)}{q^2 T\zeta\eta\rho}\left[A\zeta\eta\rho + \beta(c_1 + \hat{c_1}\xi)z_d^x m_3\eta\rho + \beta(c_1 + c_2)z_t^x m_2\rho + v_0 e^{-\tau I} + I + \beta(c_1 + \hat{c_1}\xi)z_p^x m_1\right]$$
$$- \frac{T\zeta\eta}{2\lambda}\left[\lambda(r_1 + \hat{r_1}\xi)p + (r_2 + \hat{r_2}\xi)k(\rho(\lambda - 1) - \lambda + 2)\right] \tag{28}$$

Now, set Equations (27) and (28) equal to zero and solve for I and q:

$$-\frac{D(p)\left(1 - \tau v_0 e^{-\tau I}\right)}{qT\zeta\eta\rho} = 0 \Rightarrow I = \frac{1}{\tau}\log[\tau v_0] \tag{29}$$

Then:

$$\frac{D(Ap)}{q^2 T\zeta\eta\rho}\left[A\zeta\eta\rho + \beta(c_1 + \hat{c_1}\xi)z_d^x m_3\eta\rho + \beta(c_1 + c_2\xi)z_t^x m_2\rho + v_0 e^{-\tau I} + I + \beta(c_1 + \hat{c_1}\xi)z_p^x m_1\right]$$
$$- \frac{T\zeta\eta}{2\lambda}\left[\lambda(r_1 + \hat{r_1}\xi)p + (r_2 + \hat{r_2}\xi)k(\rho(\lambda - 1) - \lambda + 2)\right] = 0$$
$$\Rightarrow q = \frac{1}{T\zeta\eta}\sqrt{\frac{2\lambda\left[D(p)\left[\begin{array}{c}A\zeta\eta\rho + \beta(c_1 + \hat{c_1}\xi)z_d^x m_3\eta\rho + \beta(c_1 + c_2)\\ \times z_t^x m_2\rho + v_0 e^{-\tau I} + I + \beta(c_1 + \hat{c_1}\xi)z_p^x m_1\end{array}\right]\right]}{\rho\left[\lambda(r_1 + \hat{r_1}\xi)p + (r_2 + \hat{r_2}\xi)k(\rho(\lambda - 1) - \lambda + 2)\right]}} \tag{30}$$

In order to prove the concavity of the required objective profit function Π; can show that the following conditions:

$$\frac{\partial^2 \Pi}{\partial I^2} = -\frac{D(p)\left(1 + \tau^2 v_0 e^{-\tau I}\right)}{qT\zeta\eta\rho} < 0 \tag{31}$$

$$\frac{\partial^2 \Pi}{\partial q^2} = -\frac{2D(p)}{q^3 T\zeta\eta\rho}\left[\begin{array}{c}A\zeta\eta\rho + \beta(c_1 + \hat{c_1}\xi)z_d^x m_3\eta\rho + \beta(c_1 + c_2)z_t^x\\ \times m_2\rho + v_0 e^{-\tau I} + I + \beta(c_1 + \hat{c_1}\xi)z_p^x m_1\end{array}\right] < 0 \tag{32}$$

$$\frac{\partial \Pi}{\partial I \partial q} = \frac{\partial \Pi}{\partial q \partial I} = \frac{D(p)\left(1 - e^{-\tau I}\tau v_0\right)}{q^2 T\zeta\eta\rho} \tag{33}$$

The Hessian matrix can be found as follows:

$$H = \left[\begin{array}{cc}\frac{\partial^2 \Pi}{\partial I^2} & \frac{\partial^2 \Pi}{\partial I \partial q}\\[6pt] \frac{\partial^2 \Pi}{\partial q \partial I} & \frac{\partial^2 \Pi}{\partial q^2}\end{array}\right] = \left[\begin{array}{cc}-\frac{D(p)\left(1 + \tau^2 v_0 e^{-\tau I}\right)}{qT\zeta\eta\rho} & \frac{D(p)\left(1 - e^{-\tau I}\tau v_0\right)}{q^2 T\zeta\eta\rho}\\[10pt] \frac{D(p)\left(1 - e^{-\tau I}\tau v_0\right)}{q^2 T\zeta\eta\rho} & -\frac{2D(p)}{q^3 T\zeta\eta\rho}\left[\begin{array}{c}A\zeta\eta\rho + \beta(c_1 + \hat{c_1}\xi)z_d^x m_3\eta\rho\\ +\beta(c_1 + c_2)z_t^x m_2\rho + v_0 e^{-\tau I}\\ +I + \beta(c_1 + \hat{c_1}\xi)z_p^x m_1\end{array}\right]\end{array}\right].$$

The determinant of $|H|$ is followed:

$$|H| = -\frac{[D(p)]^2}{q^4 T^2 \zeta^2 \eta^2 \rho^2} + \frac{2[D(p)]^2 e^{-\tau I}\tau v_0}{q^4 T^2 \zeta^2 \eta^2 \rho^2} + \frac{2[D(p)]^2 e^{-\tau I}I\tau^2 v_0}{q^4 T^2 \zeta^2 \eta^2 \rho^2} + \frac{2A[D(p)]^2 e^{-\tau I}\tau^2 v_0}{q^4 T^2 \zeta \eta \rho}$$

$$+ \frac{[D(p)]^2 e^{-2\tau I}\tau^2 v_0^2}{q^4 T^2 \zeta^2 \eta^2 \rho^2} + \frac{2[D(p)]^2 e^{-\tau I}\beta\tau^2 c_1 m_3 v_0 z_d^x}{q^4 T^2 \zeta^2 \eta \rho} + \frac{2[D(p)]^2 e^{-\tau I}\beta\tau^2 \hat{c_1}\xi m_3 v_0 z_d^x}{q^4 T^2 \zeta^2 \eta \rho}$$

$$+ \frac{2[D(p)]^2 e^{-\tau I}\beta\tau^2 c_1 m_2 v_0 z_t^x}{q^4 T^2 \zeta^2 \eta^2 \rho} + \frac{2[D(p)]^2 e^{-\tau I}\beta\tau^2 \hat{c_1}\xi m_2 v_0 z_t^x}{q^4 T^2 \zeta^2 \eta^2 \rho} + \frac{2[D(p)]^2 e^{-\tau I}\beta\tau^2 c_1 \xi m_1 v_0 z_p^x}{q^4 T^2 \zeta^2 \eta^2 \rho^2}$$

$$+ \frac{2[D(p)]^2 e^{-\tau I}\beta\tau^2 \hat{c_1}\xi m_1 v_0 z_p^x}{q^4 T^2 \zeta^2 \eta^2 \rho^2}$$

$$= \frac{[D(p)]^2 \left[\tau^2 v_0^2 e^{-2\tau I} + 2e^{-\tau I}\tau v_0 \left[\begin{array}{c} 1 + \tau I + A\tau\zeta\eta\rho + \tau\beta\hat{c_1}\xi\left(\eta\rho m_3 z_d^x + \rho m_2 z_t^x + m_1 z_p^x\right) \\ +\tau\beta\hat{c_1}\xi\left(\eta\rho m_3 z_d^x + \rho m_2 z_t^x + m_1 z_p^x\right) \end{array}\right] - 1\right]}{q^4 T^2 \zeta^2 \eta^2 \rho^2}$$

$$= \frac{[D(p)]^2 \left[\tau^2 v_0^2 e^{-2\tau I} + 2B - 1\right]}{q^4 T^2 \zeta^2 \eta^2 \rho^2} > 0$$

where $B = e^{-\tau I}\tau v_0 \left[\begin{array}{c} 1 + \tau I + A\tau\zeta\eta\rho + \tau\beta\hat{c_1}\xi\left(\eta\rho m_3 z_d^x + \rho m_2 z_t^x + m_1 z_p^x\right) \\ +\tau\beta\hat{c_1}\xi\left(\eta\rho m_3 z_d^x + \rho m_2 z_t^x + m_1 z_p^x\right) \end{array}\right] > 1.$

The behavior of the concavity for objective function $\prod$ with respect to our decision variables $\zeta, \eta, \rho, T, p, I$ and q, the algorithm is similar to Wangsa et al. [22], and was developed to draw the global maximum feasible solutions for $I^*, \zeta^*, \eta^*, \rho^*, T^*, p^*, q^*$ and $\prod^*$. $\square$

3.9. Algorithm

Step 1. First compute I from Equation (29).
Step 2. a. set $\zeta = 1$.
b. set $\eta = 1$.
c. set $\rho = 1$.
d. set $T = 1$.
e. set $p = 20$.
Step 3. Compute the optimal q^* by using Equation (30).
Step 4. After completed Step 3, compute the actual electrical power capacities.
a. Power generation
The actual capacity of the power generation $z_p^y = qT\zeta\eta\rho\,\Omega$.
If $z_p^y \leq z_p^x$ then find $q = \frac{z_p^y}{T\zeta\eta\rho\,\Omega}$ and go to Step 5.
b. Transmission substation
The actual capacity of the transmission substation $z_t^y = qT\zeta\eta\,\Omega$.
If $z_t^y \leq z_t^x$ then find $q = \frac{z_t^y}{T\zeta\eta\,\Omega}$ and go to Step 5.
c. Distribution substation
The actual capacity of the distribution substation $z_d^y = qT\zeta\,\Omega$.
If $z_t^y \leq z_t^x$ then find $q = \frac{z_d^y}{T\zeta\,\Omega}$ and go to Step 5.
Step 5. Computed $\prod$ from Equation (16).
Step 6. Set $\zeta = 1 + 1$ and repeated Step 3 to Step 5.
Step 7. If $\prod(q_\zeta^*, \zeta, \eta_\zeta, \rho_\zeta, T_\zeta, p_\zeta) \geq \prod(q_{\zeta-1}^*, \zeta - 1, \eta_{\zeta-1}, \rho_{\zeta-1}, T_{\zeta-1}, p_{\zeta-1})$ then go to Step 8. Otherwise go to Step 6.
Step 8. Set $\eta = 1 + 1$ and repeated Step 2b to Step 7.
Step 9. If $\prod(q_\eta^*, \zeta_\eta^*, \eta, \rho_\eta, T_\eta, p_\eta) \geq \prod(q_{\eta-1}^*, \zeta_{\eta-1}^*, \eta - 1, \rho_{\eta-1}, T_{\eta-1}, p_{\eta-1})$ then go to Step 9. Otherwise go to Step 8.
Step 10. Set $\rho = 1 + 1$ and repeated Step 2c to Step 9.
Step 11. If $\prod(q_\rho^*, \zeta_\rho^*, \eta_\rho^*, \rho, T_\rho, p_\rho) \geq \prod(q_{\rho-1}^*, \zeta_{\rho-1}^*, \eta_{\rho-1}, \rho - 1, T_{\eta-1}, p_{\eta-1})$ then go to Step 12. Otherwise go to Step 10.

Step 12. Set $T = 1 + 1$ and repeated Step 2d to Step 11.

Step 13. If $\prod(q_T^*, \zeta_T^*, \eta_T^*, \rho_T^*, T, p_T) \geq \prod(q_{T-1}^*, \zeta_{T-1}^*, \eta_{T-1}^*, \rho_{T-1}^*, T-1, p_{T-1})$ then go to Step 14. Otherwise go to Step 12.

Step 14. Set $p = 1 + 1$ and repeated Step 2e to Step 13.

Step 15. If $\prod(q_p^*, \zeta_p^*, \eta_p^*, \rho_p^*, T_p^*, p) \geq \prod(q_{p-1}^*, \zeta_{p-1}^*, \eta_{p-1}^*, \rho_{p-1}^*, T_{p-1}^*, p-1)$ then go to Step 16. Otherwise go to Step 14.

Step 15. If $\prod(q_p^*, \zeta_p^*, \eta_p^*, \rho_p^*, T_p^*, p^*) \geq \prod(q_{p-1}^*, \zeta_{p-1}^*, \eta_{p-1}^*, \rho_{p-1}^*, T_{p-1}^*, p-1)$ then find I^*, $\zeta^*, \eta^*, \rho^*, T^*, p^*, q^*$ and go to Step 16.

Step 16. Stop.

4. Numerical Example

In this section, used data to demonstrate the application of the model. This study considers the sustainable electrical energy supply chain inventory system in the Taiwanese electrical production industry to determine the optimal ordering quantity and total profit. The parameters in this section are assumed from a previous published paper. The data values are:

4.1. Customer Data

Let $a = 80\ units$, $b = 2\ units$, $A = \$100/unit/order$, $r_1 = \$\,0.004/unit/year$, $\hat{r}_1 = 0.002\ ton/year$, $r_2 = \$\,0.002/unit/year$, $\hat{r}_2 = 0.003\ ton/year$, $\beta = 0.01\ unit$, $\Omega = 1.2\ kVA/kWh$, $v_0 = \$7/setup$, $\xi = \$1/ton/year$ and $\tau = 0.2\ unit$.

4.2. Transmission Substation, Distribution Substation and Power Generation Data

Let $k = \$4/kWh$, $\lambda = 2\ units$, $c_1 = \$0.00011/kVA/mile$, $z_p^x = 6\ kVA$, $z_t^x = 5\ kVA$, $z_d^x = 2\ kVA$, $m_1 = 0.2\ mile$, $m_2 = 0.15\ mile$ and $m_3 = 0.1\ mile$.

Based on input data, by using Algorithm and Mathematica 9.0, the optimal solutions (see Tables 3 and 4); $\zeta^* = 2$, $\eta^* = 1$, $\rho^* = 1$, $T^* = 1$, $p^* = 22$, $D(p)^* = 36$, $I^* = 1.68236$, $q^* = 183.875$, $z_p^{y*} = 441.30$, $z_t^{y*} = 441.30$, $z_d^{y*} = 441.30$ and $\prod^* = 601.653$. Furthermore, Figures 2 and 3 for graphical representations of the total profit function is a concave with respect to feasible optimal values p^* and q^*.

Table 3. Details of the procedures for the solution.

Iteration	ζ^*	η^*	ρ^*	T^*	p^*	$D(p)^*$	I^*	q^*	z_p^{y*}	z_t^{y*}	z_d^{y*}	$\prod^*$
1	1	1	1	1	20	40	1.68236	66.6182	79.94184	79.94184	79.94184	571.609
2	2	1	1	1	20	40	1.68236	185.451	445.0824	445.0824	445.0824	593.597
3	3	1	1	1	20	40	1.68236	338.854	1219.874	1219.874	1219.874	561.851
4	4	1	1	1	20	40	1.68236	520.277	2497.329	2497.329	2497.329	496.906
5	5	1	1	1	20	40	1.68236	725.913	4355.478	4355.478	4355.478	398.489
6	6	1	1	1	20	40	1.68236	953.188	6862.953	6862.953	6862.953	264.009
7	2	2	1	1	20	40	1.68236	520.277	2497.329	2497.329	1248.664	496.906
8	2	3	1	1	20	40	1.68236	953.188	6862.953	6862.953	2287.651	264.009
9	2	1	2	1	20	40	1.68236	381.779	1832.539	916.2696	916.2696	575.894
10	2	1	1	2	20	40	1.68236	370.901	1780.324	1780.324	1780.324	537.988
11	2	1	1	1	21	38	1.68236	184.879	443.7096	443.7096	443.7096	599.611
12	2	1	1	1	22	36	1.68236	183.875	441.3000	441.3000	441.3000	601.653←
13	2	1	1	1	23	34	1.68236	182.430	437.8320	437.8320	437.8320	599.736

* The local maximum solution; ← the optimal solution.

Table 4. The results.

Decision Variables	Values
Electrical power distribution factor	2 Times
Electrical power transmission factor	1 Times
Electrical power generation	1 Times
Customer's average electricity consumption	1 Year
Retailer's price of electricity	$22/kWh.
Demand of customer	36 kWh/year
Electrical power consumption	183.875 kW
Investment for setup cost reduction	$1.68236/ production run
Energy transmitted by power generation	8.55471 kWh
Energy transmitted by transmission substation	4.277 kWh
Energy transmitted by distribution substation	6.41593 kWh
Energy consumed by customer	183.875 kWh
Actual capacity of power generation	441.3000 kVA
Actual capacity of transmission substation	441.3000 kVA
Actual capacity of distribution substation	441.3000 kVA
Power supply rate to power generation	72 kWh/year

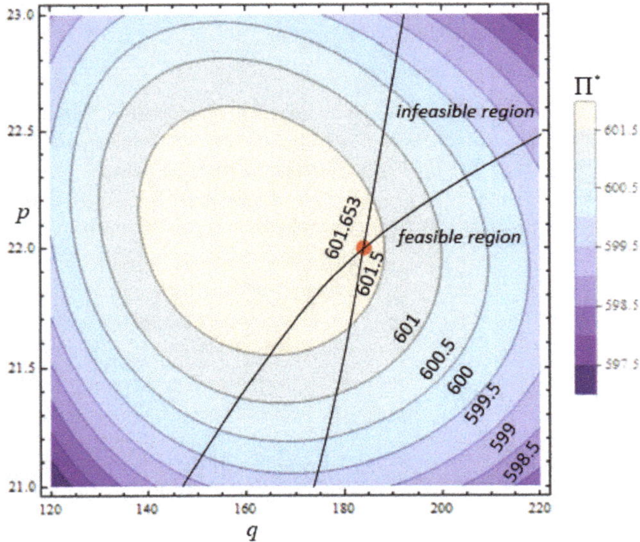

Figure 2. The contour of the objective function Π^* with respect to $\zeta = 2, \eta = 1, \rho = 1, I = 1.68236$ and $T = 1$.

From Table 3, we obtain the $\rho = 1$ batch. It is intended that the electricity produced by a power generator is 8.55471 kWh. But not all the electricity energy (8.55471 kWh) induced is transmitted at once, but in periods with 8.55471 kWh each. As the generator has a device to minimize the total energy holding cost of the electricity, for each batch, the transmission substation obtains 4.277 kWh of electricity; it then transmits in 2 batches to the distribution station at 6.41593 kWh each. This is done to minimize the transmission cost and distribution cost. Based on those results, the total electrical power consumption of customer 183.875 kW. The demand of customer is 36 kWh/year.

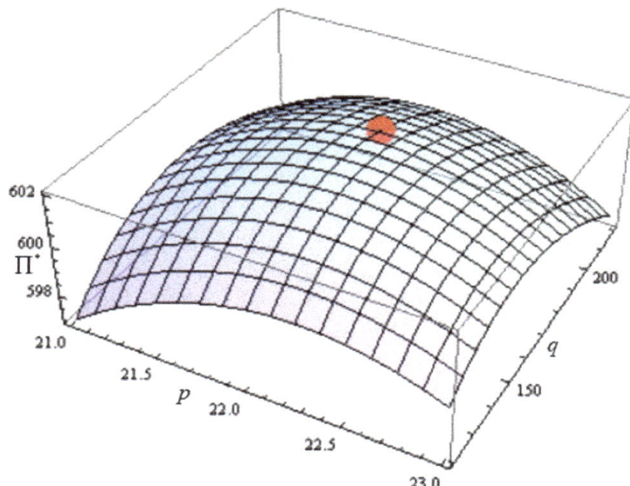

Figure 3. Concavity of $\prod^*$ with respect to p and q at $\zeta = 2$, $\eta = 1$, $\rho = 1$, $I = 1.68236$ and $T = 1$.

5. Discussion and Managerial Implication

In this section, we consider the effect of changes in the main parameters as well as summarize the results of the sensitivity analysis, which shows (see Table 5) the impact of each of the parameters which are a, b, k, λ, v_0, τ, A, $\hat{c}_1$, $\hat{r}_1$, $\hat{r}_2$ and β, respectively, on the total profit.

Table 5. Sensitivity analysis of key parameters of the model.

Parameter	Changes	ζ^*	η^*	ρ^*	T^*	p^*	$D^*(p)$	q^*	I^*	z_p^{y*}	z_t^{y*}	z_d^{y*}	$\prod^*$
	60	2	1	1	1	20	20	131.133	1.68236	314.719	314.719	314.719	287.189
	70	2	1	1	1	20	30	160.605	1.68236	385.452	385.452	385.452	439.814
a	80	2	1	1	1	22	36	183.875	1.68236	441.300	441.300	441.300	601.653
	90	2	1	1	1	24	42	206.829	1.68236	496.389	496.389	496.389	787.868
	100	2	1	1	1	27	46	228.775	1.68236	549.060	549.060	549.060	997.886
	1.8	3	1	1	1	24	36.8	353.749	1.68236	1273.49	1273.49	1273.49	643.645
	1.9	2	1	1	1	23	36.3	188.50	1.68236	452.400	452.400	452.400	641.897
b	2	2	1	1	1	22	36	183.875	1.68236	441.300	441.300	441.300	601.653
	2.1	2	1	1	1	21	35.9	179.698	1.68236	431.275	431.275	431.275	565.211
	2.2	2	1	1	1	20	36	175.934	1.68236	422.241	422.241	422.241	531.978
	2	2	1	1	1	21	38	181.449	1.68236	435.477	435.477	435.477	676.583
	3	2	1	1	1	22	36	182.249	1.68236	437.397	437.397	437.397	638.158
k	4	2	1	1	1	22	36	183.875	1.68236	441.300	441.300	441.300	601.653
	5	2	1	1	1	23	34	183.965	1.68236	441.516	441.516	441.516	565.209
	6	2	1	1	1	23	34	185.486	1.68236	445.166	445.166	445.166	530.673
	1.8	2	1	1	1	22	36	220.455	1.68236	529.092	529.092	529.092	575.565
	1.9	2	1	1	1	22	36	204.23	1.68236	490.152	490.152	490.152	590.137
λ	2	2	1	1	1	22	36	183.875	1.68236	441.300	441.300	441.300	601.653
	2.1	2	1	1	1	22	36	157.798	1.68236	378.715	378.715	378.715	609.451
	2.2	2	1	1	1	22	36	122.397	1.68236	293.752	293.752	293.752	611.235

Table 5. *Cont.*

Parameter	Changes	ζ^*	η^*	ρ^*	T^*	p^*	$D^*(p)$	q^*	I^*	z_p^{y*}	z_t^{y*}	z_d^{y*}	Π^*
v_0	5	2	1	1	1	22	36	183.81	1.53742	441.144	441.144	441.144	601.668
	6	2	1	1	1	22	36	183.81	1.61042	441.144	441.144	441.144	601.600
	7	2	1	1	1	22	36	183.875	1.68236	441.300	441.300	441.300	601.653
	8	2	1	1	1	22	36	183.906	1.75328	441.374	441.374	441.374	601.644
	9	2	1	1	1	22	36	183.937	1.82322	441.448	441.448	441.448	601.636
τ	0.18	2	1	1	1	22	36	183.945	1.28395	441.468	441.468	441.468	601.635
	0.19	2	1	1	1	22	36	183.911	1.50094	441.386	441.386	441.386	601.643
	0.2	2	1	1	1	22	36	183.875	1.68236	441.300	441.300	441.300	601.653
	0.21	2	1	1	1	22	36	183.837	1.83458	441.208	441.208	441.208	601.662
	0.22	2	1	1	1	22	36	183.797	1.96265	441.112	441.112	441.112	601.672
A	80	2	1	1	1	22	36	165.126	1.68236	396.302	396.302	396.302	606.378
	90	2	1	1	1	22	36	174.752	1.68236	419.404	419.404	419.404	603.952
	100	2	1	1	1	22	36	183.875	1.68236	441.300	441.300	441.300	601.653
	110	2	1	1	1	22	36	192.566	1.68236	462.158	462.158	462.158	599.462
	120	2	1	1	1	22	36	200.881	1.68236	482.114	482.114	482.114	597.366
$\hat{c}_1$	0.00012	2	1	1	1	22	36	183.875	1.68236	441.300	441.300	441.300	601.653
	0.0002	2	1	1	1	22	36	183.875	1.68236	441.300	441.300	441.300	601.651
	0.0003	2	1	1	1	22	36	183.875	1.68236	441.300	441.300	441.300	601.649
	0.0004	2	1	1	1	22	36	183.875	1.68236	441.300	441.300	441.300	601.647
	0.0005	2	1	1	1	22	36	183.875	1.68236	441.300	441.300	441.300	601.645
$\hat{r}_1$	0.002	2	1	1	1	22	36	183.875	1.68236	441.300	441.300	441.300	601.653
	0.003	2	1	1	1	22	36	197.606	1.68236	474.254	474.254	474.254	596.761
	0.004	2	1	1	1	22	36	210.443	1.68236	505.063	505.063	505.063	591.175
	0.005	2	1	1	1	22	36	222.541	1.68236	534.099	534.099	534.099	584.989
	0.006	2	1	1	1	22	36	234.014	1.68236	561.634	561.634	561.634	578.275
$\hat{r}_2$	0.003	2	1	1	1	22	36	183.875	1.68236	441.300	441.300	441.300	601.653
	0.004	2	1	1	1	22	36	185.165	1.68236	444.396	444.396	444.396	601.240
	0.005	2	1	1	1	22	36	186.447	1.68236	447.473	447.473	447.473	600.821
	0.006	2	1	1	1	22	36	187.719	1.68236	450.526	450.526	450.526	600.395
	0.007	2	1	1	1	22	36	188.983	1.68236	453.559	453.559	453.559	599.962
β	0.008	2	1	1	1	22	36	183.875	1.68236	441.300	441.300	441.300	601.652
	0.009	2	1	1	1	22	36	183.875	1.68236	441.300	441.300	441.300	601.652
	0.01	2	1	1	1	22	36	186.447	1.68236	441.300	441.300	441.300	601.652
	0.011	2	1	1	1	22	36	183.875	1.68236	441.300	441.300	441.300	601.652
	0.012	2	1	1	1	22	36	183.875	1.68236	441.300	441.300	441.300	601.652

Based on the computational results (Table 5), the following managerial insights can be obtained:

5.1. Impact on Demand Parameters

An increase in the value of a results in an increase in demand, which forces the retailer to increase the selling price p, fixed ζ, η, ρ, T, I, increases the actual capacity of the distribution, transmission, power generation substation and the customer's power consumption in order to increase the electrical supply chain profit. On the other hand, an increase in the value of b could reduce the demand. So, the retailer then reduces the selling price p, fixed η, ρ, T, I, reduces the actual capacity of the distribution, transmission, power generation substation and the customer's power consumption. In this case, although the demand rate could be maintained on the higher side with the high value of b. Therefore, the profit of the electrical supply chain system would be decreased.

5.2. Impact on Production Parameters

An increase in the value of λ results in an increase in production with fix at demand rates, which forces the retailer to fix the selling price p, ζ, η, ρ, T, I, decreases the actual capacity of the distribution, transmission, power generation substation and the customer's power consumption in order to increase the electrical supply chain profit. An increase in the value of k could reduce the demand. So, the retailer then increases the selling price p with fixed ζ, η, ρ, T, I, and increases an actual capacity of the distribution, transmission, power generation substation and the customer's power consumption. In this case, although the demand rate could be maintained in the less with the high value of k, customer's power consumption is higher, and actual capacity higher. Therefore, the profit of the electrical supply chain system is decreased.

5.3. Impact on Setup Cost Reduction Parameters

An increase in the value of v_0 results in an increase in setup cost with fixed at demand rates, which forces the retailer to fix the p, ζ, η, ρ, T, increases the actual capacity of the distribution, transmission, power generation substation and the customer's power consumption, increases the setup investment cost, and decreases the electrical supply chain profit. In this case, the profit of the electrical supply chain system decreased because all actual capacity and customer's power consumption with fixed selling price increase. On the other hand, an increase in the value of τ results in a decrease in setup cost with a fix at demand rates, which forces the retailer to fix the p, ζ, η, ρ, T, decreases the actual capacity of the distribution, transmission, power generation substation and the customer's power consumption, increases the setup investment cost and increases the electrical supply chain profit. In this case, the profit of the electrical supply chain system is increased because the demand rate maintained in the fixed with the high value of τ, customer's power consumption is lower, and all actual capacities are lower.

5.4. Impact on Ordering Cost

An increase in the value of A results in an increases the customer's power consumption rate with fixed demand rates; selling price increases the actual capacity of the distribution, transmission, and power generation substation. This result indicates that profit of the electrical supply chain system decreased because of higher of all actual capacity and customer's power consumption rates with fixed selling price.

5.5. Impact on Loss Factor

An increase in the value of all carbon emission parameters β, results in an no change of the customer's power consumption rate; demand rates; selling price, increases the actual capacity of the distribution, transmission, and power generation substation. This result indicates that, total profit can be unchanged because of all actual capacity and customer power consumption rates with fixed selling price are unchanged when small changes of β. Therefore, small changes of loss factor result in unchanged profit.

5.6. Impact on Carbon Emission Parameters

An increase in the value of all carbon emission parameters $\hat{c}_1$, $\hat{r}_1$ and $\hat{r}_2$ results in an increase the customer's power consumption rate with fixed demand rates; selling price increases the actual capacity of the distribution, transmission, and power generation substation. This result indicates that profit of the electrical supply chain system decreased because of higher actual capacity and customer power consumption rates at a fixed selling price.

6. Conclusions and Future Research

In this paper, a sustainable electricity supply chain mathematical model that assumes linear price-dependent customer demands where the price is a decision variable with reduction of setup cost under carbon emission, is considered. This model has been developed based on the inventory management theory, and examined how the all optimal decision variables and the total profits for sustainable electrical supply chain are affected by key parameters. Based on our computational results, it supplies managerial insights to the production system and marketing managers to help in planning a successful and sustainable electrical energy supply chain. For a future study, researchers can extend the present model to include green technology investment under carbon emissions regulation. Researchers can also study incorporating price discount strategies as well as the effect of green technology investment under carbon emissions with a multi-transmission and distribution substation.

Author Contributions: Data Curation, Software U.M. and J.-Z.W.; Formal Analysis, Investigation, Writing Original Draft Preparation, Writing-Review, Visualization and Editing U.M.; Conceptualization, Methodology, Validation A.C and J.-Z.W.; Funding Acquisition, Supervision, Project Administration J.-Z.W.

Funding: This research was funded by Ministry of Science and Technology, Taiwan, grant number MOST106-2218-E-007-024 & MOST108-2911-I-031-502; and The APC was funded by MOST106-2218-E-007-024 & MOST108-2911-I-031-502.

Acknowledgments: The authors would like to thank the editor and anonymous reviewers for their valuable and constructive comments, which have led to a significant improvement in the manuscript.

Conflicts of Interest: The authors declare no conflict of interest.

References

1. Yang, M.F. Supply chain integrated-inventory model with present value and dependent crashing cost is polynomial. *Math. Comput. Modell. Dyn.* **2010**, *51*, 802–809. [CrossRef]
2. Hoque, M.A.A. Vendor-buyer integrated-production-inventory model with normal distribution of lead time. *Int. J. Prod. Econ.* **2013**, *144*, 409–417. [CrossRef]
3. Sarkar, B.; Gupta, H.; Chaudhuri, K.S.; Goyal, S.K. An integrated-inventory model with variable lead time, defective units and delay-in-payments. *Appl. Math. Comput.* **2014**, *237*, 650–658. [CrossRef]
4. Mishra, U. Optimizing a three-rates-of-production inventory model under market selling price and advertising cost with deteriorating items. *Int. J. Manage. Sci. Eng. Manag.* **2018**, *13*, 295–305. [CrossRef]
5. Fauza, G.; Amer, Y.; Lee, S.H.; Prasetyo, H. An integrated single-vendor multi-buyer production-inventory policy for food products incorporating quality degradation. *Int. J. Prod. Econ.* **2016**, *182*, 409–417. [CrossRef]
6. Denizel, M.; Erengüç, S.; Benson, H.P. Dynamic lot-size with setup cost reduction. *Eur. J. Oper. Res.* **1997**, *100*, 537–549. [CrossRef]
7. Diaby, M. Integrated batch size and setup reduction decisions in multi-product, dynamic manufacturing environments. *Int. J. Prod. Econ.* **2000**, *67*, 219–233. [CrossRef]
8. Nyea, T.J.; Jewkes, E.M.; Dilsc, D.M. Theory and Methodology Optimal investment in setup reduction in manufacturing systems with WIP inventories. *Eur. J. Oper. Res.* **2001**, *135*, 128–141. [CrossRef]
9. Freimer, M.; Thomas, D.; Tyworth, J. The value of setup cost reduction and process improvement for the economic production quantity model with defects. *Eur. J. Oper. Res.* **2006**, *173*, 241–251. [CrossRef]
10. Huang, C.K.; Cheng, T.L.; Kao, T.C.; Goyal, S.K. An integrated inventory model involving manufacturing setup cost reduction in compound Poisson process. *Int. J. Prod. Econ.* **2011**, *49*, 1219–1228. [CrossRef]
11. Sarkar, B.; Moon, I. Improved quality, setup cost reduction, and variable backorder costs in an imperfect production process. *Int. J. Prod. Econ.* **2014**, *155*, 204–213. [CrossRef]
12. Sarkar, B.; Chaudhuri, K.S.; Moon, I. Manufacturing setup cost reduction and quality improvement for the distribution free continuous-review inventory model with a service level constraint. *J. Manuf. Syst.* **2015**, *34*, 74–82. [CrossRef]
13. Sarkar, B.; Ganguly, B.; Sarkar, M.; Pareek, S. Effect of variable transportation and carbon-emission in a three-echelon supply chain model. *Transp. Res. Part E* **2016**, *91*, 112–128. [CrossRef]
14. Sarkar, B.; Saren, S.; Sarkar, M.; Seo, Y.W. A Stackelberg Game Approach in an Integrated Inventory Model with Carbon-Emission and Setup Cost Reduction. *Sustainability* **2016**, *8*, 1244. [CrossRef]
15. Banbury, J.G. Distribution-the final link in the electricity-supply chain. *Electron. Power.* **1975**, *21*, 773–775. [CrossRef]
16. Schneider, M.; Biel, K.; Pfaller, S.; Schaede, H.; Rinderknecht, S.; Glock, C.H. Optimal sizing of electrical energy storage systems using inventory models. *Energy Procedia* **2015**, *73*, 48–58. [CrossRef]
17. Schneider, M.; Biel, K.; Pfaller, S.; Schaede, H.; Rinderknecht, S.; Glock, C.H. Using inventory models for sizing energy storage systems: An interdisciplinary approach. *J. Storage Mater.* **2016**, *8*, 339–348. [CrossRef]
18. Taylor, J.A.; Mathieu, J.L.; Callaway, D.S.; Poolla, K. Price and capacity competition in balancing markets with energy storage. *Energy Syst.* **2017**, *8*, 169–197. [CrossRef]

19. Wu, A.; Philpott, A.; Zakeri, G. Investment and generation optimization in electricity systems with intermittent supply. *Energy Syst.* **2017**, *8*, 127–147. [CrossRef]
20. Ouedraogo, N.S. Modeling sustainable long-term electricity supply-demand in Africa. *Appl. Energy* **2017**, *190*, 1047–1067. [CrossRef]
21. Wangsa, I.D.; Wee, H.M. The economical modelling of a distribution system for electricity supply chain. *Energy Syst.* **2018**. [CrossRef]
22. Wangsa, I.D.; Yang, T.M.; Wee, H.M. The Effect of Price-Dependent Demand on the Sustainable Electrical Energy Supply Chain. *Energies* **2018**, *11*, 1645. [CrossRef]
23. Hammami, R.; Nouira, I.; Frein, Y. Carbon emissions in a multi echelon production-inventory model with lead time constraints. *Int. J. Prod. Econ.* **2015**, *164*, 292–307. [CrossRef]
24. Tang, S.; Wang, W.; Yan, H.; Hao, G. Low carbon logistics: Reducing shipment frequency to cut carbon-emissions. *Int. J. Prod. Econ.* **2015**, *164*, 339–350. [CrossRef]
25. Tang, J.; Ji, S.; Jiang, L. The Design of a Sustainable Location-Routing-Inventory Model Considering Consumer Environmental Behavior. *Sustainability* **2016**, *8*, 211. [CrossRef]
26. Ouyang, L.Y.; Ho, C.H.; Su, C.H.; Yang, C.T. An integrated inventory model with capacity constraint and order size dependent trade-credit. *Comput. Ind. Eng.* **2015**, *84*, 133–143. [CrossRef]
27. Mishra, U.; Cárdenas-Barrón, L.E.; Tiwari, S.; Shaikh, A.A.; Treviño-Garza, G. An inventory model under price and stock dependent demand for controllable deterioration rate with shortages and preservation technology investment. *Ann. Oper. Res.* **2017**, *254*, 165–190. [CrossRef]
28. Mishra, U.; Tijerina-Aguilera, J.; Tiwari, S.; Cárdenas-Barrón, L.E. Retailer's Joint Ordering, Pricing, and Preservation Technology Investment Policies for a Deteriorating Item under Permissible Delay in Payments. *Math. Prob. Eng.* **2018**. [CrossRef]

Article

Optimization of Safety Stock under Controllable Production Rate and Energy Consumption in an Automated Smart Production Management

Mitali Sarkar [1] and Biswajit Sarkar [2,*]

[1] Department of Industrial Engineering, Yonsei University, 50 Yonsei-ro, Sinchon-dong, Seodaemun-gu, Seoul 03722, Korea; mitalisarkar.ms@gmail.com
[2] Department of Industrial & Management Engineering, Hanyang University, Ansan, Gyeonggi-do 155 88, Korea
* Correspondence: bsbiswajitsarkar@gmail.com; Tel.: +82-10-7498-1981

Received: 28 April 2019; Accepted: 23 May 2019; Published: 29 May 2019

Abstract: A smart production system is essential to produce complex products under the consumption of efficient energy. The main ramification of controllable production rate, amount of production size, and safety stocks is simultaneously optimized under proper utilization of energy within a smart production system with a random breakdown of spare parts. Due to the random breakdown, a greater amount of energy may be used. For this purpose, this study is concerned about the optimum safety stock level under the exact amount of energy utilization. For random breakdown, there are three cases as production inventory meets the demand without utilization of the safety stock, with using of the safety stock, and consumed the total safety stock amount and facing shortages. After the random breakdown time, the smart production system may move to an out-of-control state and may produce defective items, where the production rate of defective items is a random variable, which follows an exponential distribution. The total cost is highly nonlinear and cannot be solved by any classical optimization technique. A mathematical optimization tool is utilized to test the model. Numerical study proves that the effect of energy plays an important role for the smart manufacturing system even though random breakdowns are there. it is found that the controllable production rate under the effect of the optimum energy consumption really effects significantly in the minimization cost. It saves cost regarding the corrective and preventive maintenance cost. The amount of safety stock can have more support under the effect of optimum energy utilization. The energy can be replaced by the solar energy.

Keywords: renewable energy; smart production system; random breakdown; safety stock; controllable production rate

---

1. Introduction

Recently, the main development in the production sector is the inspection of smart manufacturing within the production context. The effect of energy is another critical milestone for the smart manufacturing system (see for reference Sarkar et al. [1]). Cárdenas-Barrón [2] introduced an optimum production strategy for those products, which have special property like as fixed lifetime but that model did not consider the machine breakdown during making the production policies. Thus, there was no assumption for using the extra energies during machine breakdown and how to reduce those energies by using renewable energy. This proposed model would like to consider this research gap within the literature to obtain the minimum total cost for the smart production system. Talizadeh et al. [3] introduced the outsourcing reworked strategy for defective products within a traditional production system, but there is not a single assumption regarding energy reduction or uses of renewable energy.

That research gap is also fulfilled by this proposed research. Wee et al. [4] introduced renewable energy within the supply chain management for the performance, limitations for their uses, and further improvement for different policies. But, there is no indication how to transfer non-renewable to renewable energy within a smart production system. Therefore, this study introduces the strategy for reduced energy consumption through renewable energy resources.

The main aim of this smart production is to make a controllable production for making more smart products than the existing one and to reduce human labor's efforts. But the main difference between an additive manufacturing and a smart manufacturing is that the additive manufacturing prefers more machinery controls than human labor, whereas the main theme of smart manufacturing systems is to make a human-factor related complex controllable management such that human-machine interactions can be used in the proper way to obtain the best complex products Kang et al. ([5]) through skilled nature of labors. Kang et al. [5] proved that the human interaction system must be controlled by a smart manufacturing system where the main performance of inspection is conducted by the three specified labors like skilled, unskilled, and semi-skilled labors. But the main issue is that they did not consider anything in the direction of energy. However, in any smart manufacturing system, the main base is smart machines and human labors and the interactions between them. Within the smart production system, there are two policies for breakdown case as no-resumption (NR) and abort/resume (AR). NR means the production may start again after repair of defective machine provided all inventories are utilized and AR means the production will start just after the repair of breakdown machine with the condition that the existing inventory should deplete to a certain amount of safety stock. Therefore, to maintain the breakdown of the system, the preventive and corrective maintenance are used. Even though both the maintenance are used, the inventory can touch the zero level without facing shortages, it may happen that inventory can go below the zero level but, it still lies in between zero and safety stock level, or sometimes it may happen that the inventory may reach to the downside of the safety stock and facing the shortages. Therefore, it can be understood that the safety stock plays an important role, whose decision is generally taken by the management sector of the smart manufacturing system. Porteus [6] first introduced the concept of improvement for the production system's quality and reduction of setup cost by using some investments. Khouja and Mehraj [7] first introduced the controllable production rate, where the production rate is variable within a certain interval of production capacity. They used that the unit production cost is dependent on a variable production rate. This is the basic idea for the controllable production rate within a traditional production model, which is the major turning of the smart production without having any consideration of energy consumption. Chung and Hou [8] extended the basic production model Khouja and Mehraj [7] with shortages, imperfect production, and the elapsed time duration of the process distribution without anything related with energy. Giri and Dohi [9] first introduced the random breakdown within a production context, where the production lot size is a decision variable and the failure rate is considered as random exponential distribution. Avancini et al. [10] defined several matrices for energy evaluation in any smart system. They explained how the intelligent network supports for energy efficiency but there is a lack of specification any production system. Because, each production system is different with each other for the purpose of energy consumption issue. They just described about the control, communication, and display with the efficient energy. Chen and Lo [11] developed warranty policy for defective products in an imperfect production system where shortages appear due to random defective products. Even though the defective products are produced, there is no concept of the optimum energy consumption. Nižetić et al. [12] introduced several smart technologies to manage wastes through the efficient energies. They used sustainable resources to reduce the global energy but it has lack of concept within the smart production system. Specifically, the effect of efficient energy was not considered by any researcher yet. An economic production model for imperfect production system was developed by Sarkar and Moon [13]. They considered defective products when the system moves to *out-of-control* state from the *in-control* state. They did not consider any type of preventive or corrective maintenance. They just considered a traditional production system with *out-of-control* movement without any concept of energy. A two-echelon supply chain management is explained with probabilistic deterioriation in Sarkar [14]. The aim of the supply chain is to reduce the total cost but with a traditional production system. Any energy consumption is not considered within the whole supply chain model. For any smart production system, smart optimum energy and finally smart optimum energy system are very much essential to make any grid system with the help of grid system, the smart machine can be worked properly within several smart production systems, but Lund et al. [15] described only the explanation of grid system without modelling any specific smart production system. Generally, the smart manufacturing utilizes smart integrating intelligent manufacturing machines to produce smart products. The process is totally controlled though automated machines Edgar and Pistikopoulos [16]. Sana et al. [17] discussed that those defective products can be sold by reduced price. However, none of the authors did consider energy consumption in any production context. Louly and Dolgui [18] derived the optimum safety stock level in an assembly production system but without any concept of energy. Jaber et al. [19] were the pioneer researcher about the incorporation the entropy/heat effect in any inventory model but no energy effect. Sarkar et al. [20] explained the joint effect of failure rate, safety stock, and production lot in a basic production mode under a corrective maintenance and preventive maintenance. They proved that safety stock production model converges always over the non-safety stock mode but without any concept of energy or heat effect. Sana [21] developed a production model with machine breakdown under both types of maintenance where the shifting mode for machine follows an exponential distribution. Sarkar et al. [22] developed a closed-loop supply chain model with environmental issues. In this model remanufacturing of returned products are considered. But, any energy issue were not take into consideration. A supply chain model for complementary products was proposed by Sarkar et al. [23] without any energy effect, but the variable production rate and time dependent holding costs are applied. A logistic model was developed by Sarkar et al. [24], where the optimal cash flow for a smart production system is described. In this model carbon foot print and carbon emission costs were considered witout any energy consideration. Cárdenas-Barrón et al. [25] developed the benefit of multi-shipment in an economic manufacturing model without having any concept of energy matter. Giotitsa et al. [26] strives an appropriate approach for production of energy and its distribution. Tayyab and Sarkar [27] were the pioneer about incorpration random backorder rate within a multi-stage production system. Omair and Sarkar [28] introduced some sustainability issues within the production context without having any concept of energy consideration. Kim and Sarkar [29] explained a multi-stage cleaner production model, but they did not consider any idea about renewable energy effect within the multi-stage production system. They consider production system moves to out-of-control state even though they did not consider machine breakdown only, they just assumed the production system moves out-of-control state from in-control state due to labor strickes and machine breakdown without calculating breakdowns within inventory calculation. In this same direction, Moon et al. [30] approached a continous review model with service level and variable lead time without any smart production consideration. A two-echelon supply chain model with quality improvement and setup cost reduction in a tradition production system was considered by Sarkar et al. [31], but production rate is constant though it moves to *out-of-control* state. A supply chain model Kim and Sarkar [32] was developed with stochastic lead time and transportation discount policy without any maintaince policy. A study on setup cost reduction and quality improvement of manufacturing system was proposed by Majumder et al. [33] within a traditional production system. For an imprfect production system, lost sales reduction and quality improvement was taken account in fuzzy environment by Soni et al. [34], but still they did not think about the smart production system. Sarkar et al. [35] explained about a two-echelon supply chain model with setup time reduction and effect of safety factor in a traditional system. An integrated inventory model was developed by Dey et al. [36] on variable safety fctor and setup cost reduction. But for all of these models, maintainenece costs along with the energy costs are not considered for any stage of the supply chain or the production. Sarkar et al. [37] explained only out-of-control machines but did not consider the breakdown effects on the inventory of the production system. Biel and Glock [38] discussed an efficient way to use waste energy within a two-stage controllable production system. Kim et al. [39] developed a cleaner multi-stage production system with a random backorder rate without any sustainable issue or energy issue. Unver and Kara [40] developed the efficiency of energy consumption within a traditional process with the optimum energy consumption. They did not consider any maintenance or any controllable smart production system. Kumar et al. [41] maintained the efficient balance level within a production system. They calculated the amount of wastes in a downstream production system and they obtained several strategies to reduce those wastes. Morato et al. [42] did the reverse way of energy issue. They calculated the amount of energy during the production system for the agricultural products. Ahmed and Sarkar [43] introduced sustainable framework and an energy issue for the case of biofuel production. Darom et al. [44] developed two serial supply chain disruption recoveries with controlled safety-stock and constant production rate. Assid et al. [45] introduced an unreliable hybrid manufacturing for production and setup cost controlling. They used demand of new and remanufacturing products, which are reworked but they did not mention the time of rework. [46] introduced an unique way to decide the exact time of rework within a multi-stage single cycle and multi-stage multi-cycle production system even though Sarkar [46] did not take into account of energy cost and any controllable production system. Utilizing photovoltaic technology, the optimum amount of energy consumption within a production buffer stock was developed by Caro-Ruiz et al. [47]. They proved that the coordination is possible for production system with a energy factory with the optimum consumption. Finally, one can obtain the idea about the automation policy in a smart production system, where machinery systems are automatic and they are controlled by several automated system of energy (Dincer and Ezzat [48], Dincer and Al-Zareer [49], Dincer and Rosen [50], Dincer and Rosen [51], and Dincer and Bicer [52]). However, if the automated system is utilized in several research models (Lu et al. [53], Kazemi et al. [54], Kulczek [55], Harris et al. [56], Bruni et al. [57], Dehning et al. [58], and Nordborg et al. [59]), the optimum renewable energy under production maintenance is not considered. The main research gap regarding maintenance is not considered as it was assumed that smart production system always gives perfect smart products without any breakdown and without any maintenance (Khalil et al. [60], Keen et al. [61], Chen et al. [62], and Liang et al. [63]).

Still, there is a research gap in the direction of smart production direction with random breakdown under the proper application of corrective and preventive maintenance where the main task is to control used energy within the smart production system. This proposed model has solved this research gap for this matter. The Table 1 shows the research gaps of the model.

Table 1. Research contribution by several authors.

Author(s)	Energy	Smart Production System	Breakdown	Safety Stock	Production Rate	Maintenance
Sarkar et al. [1]	Consumption	Smart	Random	NA	Variable	NA
Kang et al. [2]	NA	Smart	NA	NA	Constant	NA
Porteus [3]	NA	Traditional	NA	NA	Constant	NA
Khouja and Mehraj [4]	NA	Traditional	NA	NA	Variable	NA
Giri and Dohi [6]	NA	Traditinal	Random	NA	Constant	CM and PM
Chenand Lo [7]	NA	Traditional	NA	NA	Constant	NA
Sana et al. [8]	NA	Traditional	Random	NA	Constant	CM and PM
Louly and Dolgui [9]	NA	Traditional	NA	Variable	Constant	CM and PM
Sarkar et al. [11]	NA	Traditional	Random	Variable	Constant	CM and PM
This paper	Consumption	Smart	Random	Variable	Variable	CM and PM

NA indicates that is not applicable for that paper. CM and PM stand for corrective and preventive maintenance, respectively.

2. Problem Definition, Notation, and Assumptions

In this section, problem definition, assumptions and notation are given.

2.1. Problem Definition

It is a smart production system, random breakdown and defective production exists under controllable energy consumption and production rate, then how safety stock can be maintained within the production with the minimum cost. The unit production cost is dependent on controllable production rate and efficient amount of energy. Due to the random breakdown, the corrective and preservative maintenance are considered. As random breakdown is an unusual event and similarly amount of energy is used as extra amount of energy during corrective maintenance of smart machines. There are several cases, where the existing inventory's position is before safety stock or after safety stock amount. The controllable production rate must vary within proper limit of production capacity of the machines such that the optimum amount of energy can be used. Finally, the aim is to minimize the optimum cost under the optimum production lot, safety stock and the production rate.

2.2. Assumptions

The following assumptions are considered to develop this model.

1. A smart production management is considered for single-type of products production under a controllable production rate and the optimum energy consumption.
2. During a long-run production system, the random breakdown may occur. Based on it, the model considers three cases as (i) the inventory just reaches to zero but no shortage appears (ii) the inventory may exceed the zero axis level but less than the safety stock level (iii) the inventory may cross the safety stock level and faces shortages.
3. Regular and emergency maintenance are considered under the effect of energy with two specified probability distribution functions.
4. Random breakdown may be at a random time, which follows an exponential distribution.
5. The effect of energy is considered for all possible position with the optimum amount of energy consumption.
6. Safety stock, production rate, and production quality are considered as decision variable where production rate may vary within the range of $[p_{min}, p_{max}]$. The optimum value of production rate must lie within the interval.
7. Shortages appear if the inventory crosses the safety stock level. If shortages appear, it is considered fully backordered.

2.3. Notation

Table 2 represents the notation of this model.

Table 2. Notation for parameters and variables.

Decision variables	
P	production rate (units/time)
Q	production lot size (units)
M_s	safety stock (units)
Random variable	
ρ	random variable of shifting from in-control-state to an out-of-control state
Parameters	
D	demand rate (units)
X	non-negative random variable denoting time
$F_X(t)$	failure time-distribution of X with probability density function $f_X(t) = \dfrac{d}{dt}(F_X(t))$
$A_1(S_1)$	corrective(emergency) repair time distribution with probability density function a_1 (S_1) with finite mean $\dfrac{1}{\mu_1}(>0)$
$A_2(S_2)$	preventive(regular) repair time distribution with probability density function a_2 (S_2) with finite mean $\dfrac{1}{\mu_2}(>0)$
A_0	setup cost per setup for the smart manufacturing system
A_0'	energy cost for setup the production system
C_{CRC}	corrective repair cost per unit time
C_{CRC}'	energy cost for corrective maintenance per unit time
C_{PRC}	preventive repair cost per unit time
C_{PRC}'	preventive maintenance energy cost per unit time
C_{hold}	holding cost per unit per unit time
C_{hold}'	energy cost to hold all products per unit per unit time
C_{shor}	shortage cost per unit product
C_{re}	rework cost per unit of defective item
C_{re}'	energy cost per unit to rework defective items
$h(\rho)$	probability distribution function of the shift time distribution
β	proportion of defective items produced in the out-of control state where $0 < \beta < 1$
T	cycle length of production-inventory system

3. Mathematical Model

A basic smart production is taken to produce a single-type of smart products. Even though, it is a smart production system, the machine failure exists and defective products produced. During long-run, the smart production system may move to out-of-control state from in-control state due to semi-skilled laborer's issues or machinery problems. The smart production begins to produce smart products at time $t = 0$ with a controllable production rate ρ and continues until the time t_1, when the inventory touches the maximum holding if there does not exist any smart machine breakdown. But the random breakdown occurs within this smart production system. Due to the random breakdown there are some possibilities like the preventive or corrective maintenance's time with efficient energy is less than or greater than expected time. Thus, shortages may occur. The following cases under efficient time can be founded as follows:

Case I: Inventory position due to preventive maintenance

If the preventive maintenance time under the efficient energy is less than or equal to $\dfrac{Q(P-D)}{PD}$, then amount of inventory is (see Figure 1)

$$I_1 = \int_0^{Q(P-D)/(PD)} \left[\frac{(P-D)Q^2}{2PD} + \frac{M_sQ}{D} \right] dA_1(S_1). \tag{1}$$

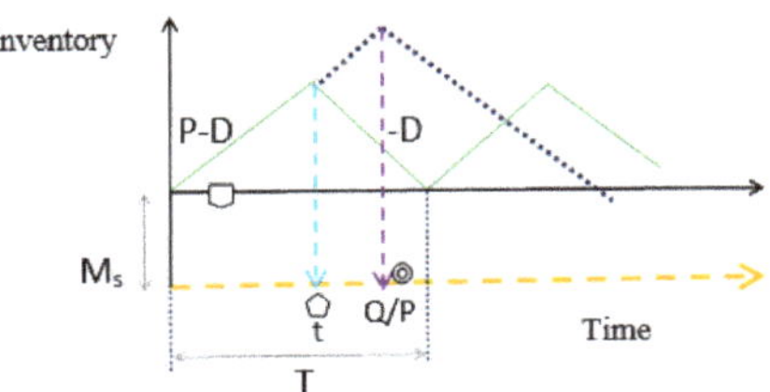

○ Starting point of corrective maintenance
◉ Ending point of corrective maintenance
▢ System moves to *out-of-control* state from *in-control* state

Figure 1. Inventory when the preventive maintenance time interval is in $\left[0, \dfrac{Q(P-D)}{PD}\right]$.

But, if the preventive maintenance time belongs to $\left[\dfrac{Q(P-D)}{PD}, \dfrac{(M_s + \dfrac{Q(P-D)}{P})}{D}\right]$, then the inventory may reach lower than the safety stock level but still there is no shortage as the inventory level is above the zero level, then the amount of inventory is (refer to Figure 2)

$$I_2 = \int_{Q(P-D)/(PD)}^{(M_s + Q(P-D)/P)/D} \left[\frac{(P-D)Q^2}{2PD} + \frac{PM_sS_1}{(P-D)} - \frac{P(S_1D - (P-D)Q/P)^2}{2D(P-D)}\right] dA_1(S_1). \quad (2)$$

Finally, if the preventive maintenance time belongs to $\left[\dfrac{(M_s + \dfrac{Q(P-D)}{P})}{D}, \infty\right)$, then the shortage arises due to the inventory level reaches below the zero level, then the total amount of inventory is (see Figure 3)

$$I_3 = \int_{(M_s + Q(P-D)/P)/D}^{\infty} \left[\frac{(P-D)Q^2}{2PD} + \frac{M_sQ}{P} + \frac{PM_s^2}{2D(P-D)})\right] dA_s(S_1), \quad (3)$$

and the inventory due to shortage for the preventive maintenance is

$$I_4 = \int_{(M_s + Q(P-D)/P)/D}^{\infty} \left[S_1 - (M_s + Q(P-D)/P)/D\right] dA_s(S_1). \quad (4)$$

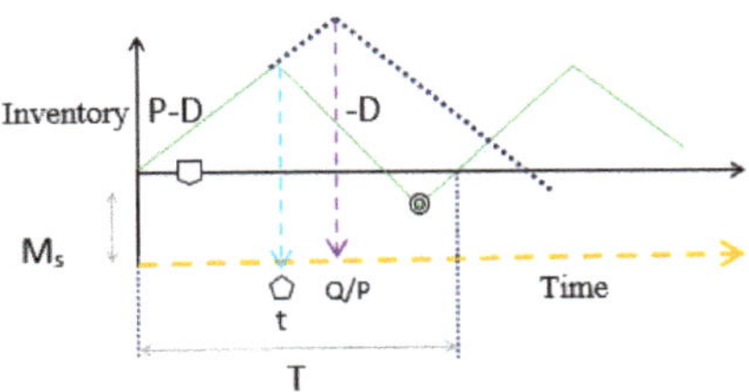

○ Starting point of corrective maintenance
◉ Ending point of corrective maintenance
▢ System moves to *out-of-control* state from *in-control* state

Figure 2. Inventory when the preventive maintenance time interval is in $\left[\dfrac{Q(P-D)}{PD}, \dfrac{(M_s + \dfrac{Q(P-D)}{P})}{D}\right]$.

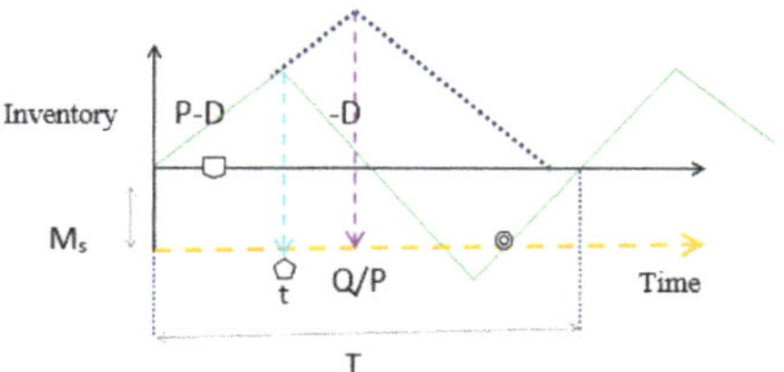

Figure 3. Inventory when preventive maintenance time interval $\left[\dfrac{(M_s + \dfrac{Q(P-D)}{P})}{D}, \infty\right)$.

Now, the inventory related with corrected maintenance can be done in case II.

Case II: Inventory position due to the corrective maintenance

The corrective maintenance is conducted when there in any urgent mishap and to save the production system, it is needed to maintenance with cost and more energy investment. Thus, one can calculate the amount of inventory position based on random time t for corrective maintenance as at random time $t = \rho$, the process moves to out-of-control state from in-control state. Similar of Case I, there are three cases when the corrective maintenance time belongs to $\left[0, \dfrac{t(P-D)}{D}\right]$, the inventory position is within the general range of inventory above zero level, or the maintenance time belongs to $\left[\dfrac{t(P-D)}{D}, \dfrac{(M_s + t(P-D))}{D}\right]$, the position of inventory should lie in between zero level and safety stock level without having shortages, but if the time for maintenance belongs to $\left[\dfrac{(M_s + t(P-D))}{D}, \infty\right)$, then inventory level reaches below the safety stock level, with having shortages, then the inventory positions, respectively are (see Figures 4–6)

$$I_5 = \int_0^{t(P-D)/D} \left[\frac{(P-D)}{2D}Pt^2 + \frac{M_s Pt}{D}\right] dA_2(S_2),\tag{5}$$

$$I_6 = \int_{t(P-D)/D}^{(M_s + t(P-D))/D} \left[\frac{(P-D)Pt^2}{2D} + \frac{PM_s S_1}{(P-D)} - \frac{P(S_1 D - (P-D)t)^2}{2D(P-D)}\right] dA_2(S_2),\tag{6}$$

$$I_7 = \int_{(M_s + t(P-D))/D}^{\infty} \left[\frac{(P-D)Pt^2}{2D} + \frac{M_s Pt}{D} + \frac{PM_s^2}{2D(P-D)}\right] dA_2(S_2),\tag{7}$$

And

$$I_8 = \int_{(M_s + t(P-D)/P)/D}^{\infty} \left[S_2 - (M_s + (P-D)t)/D\right] dA_2(S_2).\tag{8}$$

Now, one can find the expected number of defective items as the production system at random $t = \rho$ moves to out-of-control state. Therefore, the expected number of defective items are

$$I_9 = P\beta \left[\int_0^{Q/P} \int_0^t (t - P)h(\rho)dPdF_x(t) + \int_0^{Q/P} \left(\frac{Q}{P} - \rho\right)h(\rho)d\rho d\overline{F}_X\left(\frac{Q}{P}\right)\right],\tag{9}$$

where $h(\rho)$ is the probability distribution of the shifting of production system from in-control state to out-of-control state and it is given by an exponential function as follows:

$$h(\rho) = ke^{-k\rho},$$ (10)

where $\dfrac{1}{k}$ is the mean. Now other production distribution functions are given as follows:

$$F(X) \;=\; 1 - e^{-t},$$ (11)

which gives

$$\overline{F}(X) \;=\; e^{-t}.$$ (12)

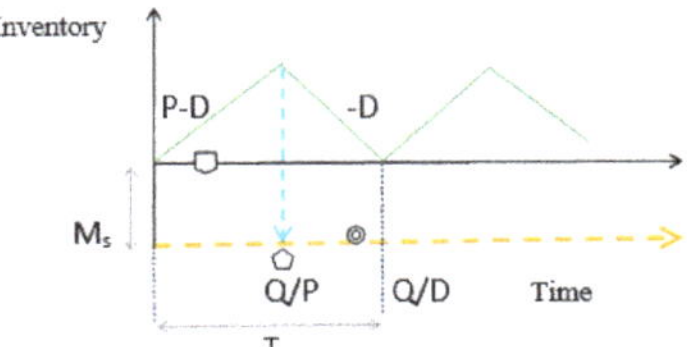

Figure 4. Inventory when the corrective maintenance time interval is in $\left[0, \dfrac{t(P - D)}{D}\right]$.

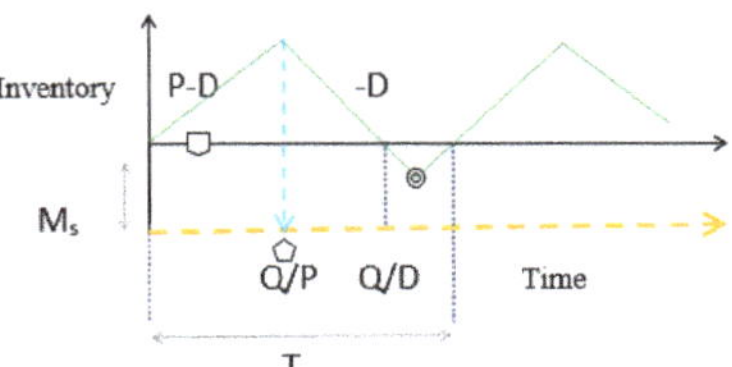

Figure 5. Inventory when the corrective maintenance time interval is in $\left[\dfrac{t(P - D)}{D}, \dfrac{(M_s + t(P - D))}{D}\right]$.

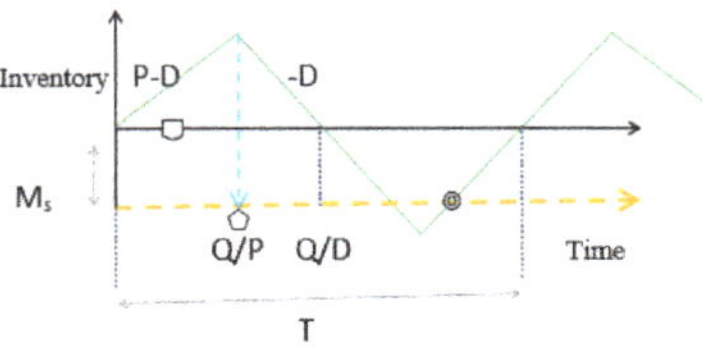

Figure 6. Inventory when the corrective maintenance time interval is in $\left[\dfrac{(M_s + t(P - D))}{D}, \infty\right)$.

The corrective and preventive repair time are given by

$$A_1(S_1) = 1 - e^{-\mu_1 S_1}, \tag{13}$$

$$A_2(S_2) = 1 - e^{-\mu_2 S_2}. \tag{14}$$

As it is a smart production system, thus the controllable unit production cost is considered and is as follows:

$$C_P = \frac{A + A'}{P} + (\gamma + \gamma')P, \tag{15}$$

where A and γ are the scaling parameters related with production cost and A' and γ' are the scaling parameters related with the energy cost for the production.

Now, the different types of costs can be calculated for the production system as follows:

As this is a smart production system, thus the setup cost acts a crucial role within the total cost of the system. To install smart machine and to maintain its reliability, the huge amount of funding support is needed. The management has to decide the whole production system will be with smart machines or partial with smart machines. For this proposed study, the management considers fully automated smart machines. Thus, they may not need any other continuous or discrete investments to continue the production. As the smart machines are used, the optimum energy is utilized for the production system. Thus, the initial setup is needed for any production system. For this purpose some constant cost along with the energy cost which is used to setup (S_{et}) the system are considered as

$$S_{et} = A_0 + A'_0. \tag{16}$$

For two types of maintenance different costs are considered here. For corrective and preventive both maintenance, energy cost is essential cost. Thus, the cost corrective and preventive maintenance can be written respectively, as

$$C_{orrect} = (C_{CRC} + C'_{CRC}) \left[\int_0^{Q/P} \left[\int_0^\infty S_2 dA_s(S_2) dF_X(t) \right] \right], \tag{17}$$

and

$$C_{preven} = (C_{PRC} + C'_{PRC}) \left[\int_0^\infty S_1 dA_s(S_1) d\overline{F}_X(Q/P) \right]. \tag{18}$$

The total inventory for both corrective and preventive maintenance are already shown. Here holding cost for all types of inventory is considered as same. To hold any inventory, its relevant energy costs are also needed. Thus, the total holding cost and its energy cost can be written as

$$(C_{hold} + C'_{hold}) \left[\int_{Q/P}^\infty (I_1 + I_2 + I_3) dF_X(t) \right] + (C_{hold} + C'_{hold}) \left[\int_0^{Q/P} (I_5 + I_6 + I_7) dF_X(t) \right]. \tag{19}$$

Rework is allowed for defective items. The inventory for rework is I_9. The rework cost and its energy cost can be written as

$$(C_{re} + C'_{re}) I_9. \tag{20}$$

Therefore, the expected total cost is given by

$$
\begin{aligned}
TC(Q, M_s, P) \;=\;& \int_0^\infty E[\text{Total cost of a cycle}|X = t]dF_X(t) \\
=\;& (A_0 + A_0') + (C_{CRC} + C_{CRC}')\left[\int_0^{Q/P}\left[\int_0^\infty S_2 dA_s(S_2)dF_X(t)\right]\right] \\
+\;& (C_{PRC} + C_{PRC}')\left[\int_0^\infty S_1 dA_s(S_1)d\overline{F}_X(Q/P)\right] + (C_P + C_P') \\
+\;& (C_{hold} + C_{hold}')\left[\int_{Q/P}^\infty (I_1 + I_2 + I_3)dF_X(t)\right] + (C_{hold} + C_{hold}')\left[\int_0^{Q/P}(I_5 + I_6 + I_7)dF_X(t)\right] \\
+\;& DC_{shor}\left[\int_{Q/P}^\infty I_4 dF_X(t) + \int_0^{Q/P} I_8 dF_X(t)\right] + (C_{re} + C_{re}')I_9,
\end{aligned}
\tag{21}
$$

and the production cycle time is given by

$$
\begin{aligned}
T(Q, M_s, P) \;=\;& \int_{Q/P}^\infty \left[\int_0^{Q(P-D)/PD} \frac{Q}{D}dA_1(S_1) + \frac{P}{P-D}\int_{Q(P-D)/PD}^{(M_s+Q(P-D)/P)/D} S_1 dA_1(S_1)\right. \\
+\;& \left.\int_{(M_s+Q(P-D)/P)/D}^\infty \left(\frac{Q}{P} + S_1 + \frac{M_s}{P-D}\right)dA_1(S_1)\right]dF_X(t) \\
+\;& \int_0^{Q/P}\left[\int_0^{t(P-D)/D} \frac{Pt}{D}dA_2(S_2) + \frac{P}{P-D}\int_{t(P-D)/D}^{(M_s+t(P-D))/D} S_2 dA_2(S_2)\right. \\
+\;& \left.\int_{(M_s+t(P-D))/D}^\infty \left(t + S_2 + \frac{M_s}{(P-D)}\right)dA_2(S_2)\right]dF_X(t).
\end{aligned}
\tag{22}
$$

Finally, using the renewal reward theorem, the expected average total cost per cycle is given by

$$
\begin{aligned}
ETC(Q, M_s, P) \;=\;& \lim_{t\to\infty} \frac{E[\text{Total cost on } (0,t)]}{t} \\
=\;& \frac{TC(Q, M_s, P)}{T(Q, M_s, P)}.
\end{aligned}
\tag{23}
$$

Equation (18) has to be optimized. This is a highly non-linear equation. Thus, it is solved with numerical experiments. The next section gives results for it.

4. Numerical Experiment

This section consists of an illustrative numerical example, sensitivity analysis, and managerial insights of the model.

4.1. Numerical Example

The revised data for numerical experiment is taken from Sarkar et al. [20] and provided in Table 3 as follows:

Table 3. Input parameters of the model.

$D = 400$ units	$A_0 = \$250$	$C_{CRC} = \$0.40/\text{unit time}$	$C_{PRC} = \$3/\text{unit time}$
$C_{re} = \$1.8/\text{item}$	$\beta = 0.2$	$C_{hold} = \$0.09/\text{unit/unit time}$	$C_{shor} = \$15/\text{item}$
$\mu_1 = 2$	$\mu_2 = 10$	$\gamma = 0.018$	$C_{PRC}' = \$2/\text{unit time}$
$A = 60$	$C_{CRC}' = \$0.10$	$A' = 5$	$C_{hold}' = \$0.01/\text{unit/unit time}$
$k = 1.2$	$A_0' = \$50$	$C_{re}' = \$0.2/\text{item}$	$\gamma' = 0.002$

The optimal solutions of the numerical example can be written as $P = 6108.89$ units/time; $Q = 2202.34$ units; $M_s = 26.34$ units; $ETC = \$343.63$.

4.2. Sensitivity Analysis

This subsection consists of the sensitivity analysis of the key parameters of the model.

From Table 4 the sensitivity analysis of key parameters can be described as follows:

(1) If the setup increased, the total cost of the smart production system increased. But wit the increases of 50% setup cost, the total cost increased only 8.62%. Thus, one can conclude that the setup cost was not effecting significantly the total cost of the production system. However as it was a smart production system, there were several smart technology and smart machinery systems and those were involved within this setup cost. Therefore, through the total cost only increased 8.62% for a 50% increase in the setup cost, still, it was a valuable and significant cost for any traditional or the smart production system.

(2) The defective cost was most sensitive within all costs of the smart production system. For negative and positive charge of the defective cost, the total cost changed 19.46% and in both directions, it was the same change. Thus, it can be concluded that the value of the rework cost follows an equilibrium position, which was the main theme for steady state of any production system. As this cost effects more, the automation policy can be used where the defective productions are inspected by a machine not by a human labor. Thus, the probability of defective products reduced and along with the defective cost and corresponding rework cost was reduced.

(3) The cost of corrective and preventive maintenance were much less sensitive with the total cost compared to the other costs. As the breakdown was random, thus the preventive and corrective maintenance should be least sensitive among all cost parameters. The main reason behind it is that the production system is a smart production system. Thus, the probability of breakdown is very less. Therefore, these costs were less sensitive. If there was a breakdown then the importance of corrective maintenance increased, but still the effect of the breakdown was not more than the regular breakdown event as the production rate was controllable and before moving to out-of-control, the management reduced the production rate. Thus, the amount of loss was much less. But, still the corrective maintenance played an important role on that time. Thus, if a breakdown occurs, then the setup cost, defective cost, corrective maintenance cost, holding cost, and shortage cost will increase significantly, whereas the preventive maintenance cost will be reduced in the next cycle more as the huge corrective maintenance is already used during maintenance.

(4) The holding cost of the smart production is very much sensitive with the total cost of the system. For negative change of the holding cost, the total cost is changed more than the positive change of it. Therefore, it can be concluded that the holding cost does not follow the equilibrium positions like the rework cost. However, the industry should try to reduce the holding cost for maintaining the minimum cost of the system. For the long term strategy to reduce holding cost, several policies should be adopted by the management such that it can reduce the total cost again.

(5) As it is a smart manufacturing system and due to the random breakdown pattern, any time the inventory level can go below the safety stock level. Therefore, the shortage cost is almost equally sensitive like other costs. Just like the holding cost, the shortage cost is also the most sensitive in the negative change than the positive change of the total cost.

Table 4. Sensitivity analysis for expected total cost.

Parameters	Changes (in %)	Changes of *ETC* (in %)	Parameters	Changes (in %)	Changes of *ETC* (in %)
A_0	-50	-10.66	C_{CRC}	-50	-0.0024
	-25	-4.97		-25	-0.0012
	$+25$	$+4.49$		$+25$	$+0.0012$
	$+50$	$+8.62$		$+50$	$+0.0024$
C_{re}	-50	-19.46	C_{hold}	-50	-17.08
	-25	-9.73		-25	-7.78
	$+25$	$+9.73$		$+25$	$+6.74$
	$+50$	$+19.46$		$+50$	$+12.68$
C_{PRC}	-50	-0.0109	C_{shor}	-50	-3.99
	-25	-0.0055		-25	-1.66
	$+25$	$+0.0055$		$+25$	$+1.29$
	$+50$	$+0.0109$		$+50$	$+2.35$

4.3. Managerial Insights

This study reveals a strong recommendation how the industry will manage the situation of the breakdown within the framework of smart production system, where generally breakdown is not expected. However, the management can decide the optimum production rate easily, based on the situation of huge or less products necessity. They can increase or decrease the production rate based on system movement from in-control to out-of-control movement. For that case, the chances of defective item production will be reduced.

A smart production system was considered for smart products. The effectiveness of energy should be taken care as major tasks are done by smart machines. Thus, the management can get the proper amount of safety stock of products with the optimum energy level. As this was a smart system, thus the rate of breakdown generally less than the traditional production system. Therefore, the wastage of energy can be reduced easily due to less number of breakdown. The management should take care about the corrective and preventive maintenance always.

As this is a smart production system with all smart machines, it is always needed more skilled labors than unskilled labors. The skilled labors will get benefits as their workloads will be reduced. Therefore, all of production and maintenance staff will be happy to accommodate this type of production system.

5. Conclusions

The study was conducted to obtain the optimum controllable production rate, the amount of the production quality and finally the amount of safety stock during the random machine breakdown under optimum energy consumption within the framework of smart production management. The total cost was minimized under the optimum energy consumption for both maintenance, corrective and preventive under the random breakdown. The variable production rate was varied within the interval of minimum and maximum range of production rate. Due to random breakdown the amount of energy was used in high rate but due to controllable production it was optimized. For a controllable production rate, the optimum safety stock is needed from the management as demand rate is constant and holding cost is comparatively low. But, due to high shortage cost, the management aimed to fulfill the demand by the optimum production quantity immediately. Thus, the controllable production rate was the best fit strategy for any smart production system. The main finding of this model was with respect to the optimum energy, the safety stock, production quantity, and production rate were optimized to obtain the minimum total cost. Sarkar et al. [1], Talizadeh et al. [3], and Sana et al. [17] considered only fixed traditional production system for complex products. But in reality, the complex products or smart technological products are easy to produce in a smart production system with a controllable production rate and proper safety stock under the optimum energy. This study fulfilled this specific research gap. Sarkar et al. [20] explained the similar breakdown in a traditional production system but this model extended that in a smart production system with the effect of controllable production rate and smart energy. Finally, the proposed model obtained the optimum cost at the optimum values of the decision variables. The main limitation of the model, the setup cost, is constant even though the random breakdowns are there. Therefore, the model can be extended with variable setup cost of the model under the similar conditions. The model can be further extended for multi-stage smart production system with random breakdowns.

Author Contributions: Data curation, software, formal analysis, investigation, writing—original draft preparation, M.S.; conceptualization, methodology, validation, visualization, funding acquisition, supervision, project administration, writing—review and editing, B.S.

Funding: The research was funded by the research fund of Hanyang University (HY-2018-F), (Project number 201800000001370).

Acknowledgments: This work was supported by the research fund of Hanyang University (HY-2018-F), (Project number 201800000001370).

Conflicts of Interest: The authors declare no conflict of interest.

References

1. Sarkar, M.; Sarkar, B.; Iqbal, M.W. Effect of energy and failure rate in a multi-item smart production system. *Energy* **2018**, *11*, 2958. [CrossRef]
2. Cárdenas-Barrón, L.E. Optimal production policy with shelf-life including shortages: A comment. *J. Oper. Res. Soc.* **2006**, *57*, 1499–1500. [CrossRef]
3. Taleizadeh, A.A.; Sari-Khanbaglo, M.P.; Cárdenas-Barrón, L.E. Outsourcing rework of imperfect items in the economic production quantity (EPQ) inventory model with backordered demand. *IEEE Trans. Syst. Man Cybern. Syst.* **2017**, *99*, 1–12. [CrossRef]
4. Wee, H.M.; Yang, W.H.; Chou, C.W.; Padilan, M.V. Renewable energy supply chains, performance, application barriers, and strategies for further development. *Renew. Sustain. Energy Rev.* **2012**, *16*, 5451–5465. [CrossRef]
5. Kang, C.W.; Ramzan, M.B.; Sarkar, B; Imran, M. Effect of inspection performance in smart manufacturing system based on human quality control system. *Int. J. Adv. Manuf. Technol.* **2018**, *94*, 4351–4364. [CrossRef]
6. Porteus, E.L. Optimal lot sizing, process quality improvement and setup cost reduction. *Oper. Res.* **1986**, *34*, 137–144. [CrossRef]
7. Khouja, M.; Mehrez, A. Economic Production Lot Size Model with Variable Production Rate and Imperfect Quality. *J. Oper. Res. Soc.* **1994**, *45*, 1405–1417. [CrossRef]
8. Chung, K.J.; Hou, K.L. An optimal production run time with imperfect production processes and allowable shortages. *Comput. Oper. Res.* **2003**, *30*, 483–490. [CrossRef]
9. Giri, B.C.; Dohi, T. Exact formulation of stochastic EMQ model for an unreliable production system. *J. Oper. Res. Soc.* **2005**, *56*, 563–575. [CrossRef]
10. Avancini, D.B.; Rodrigues, J.J.P.C.; Martins, S.G.B.; Rabêlo, R.A.L.; Al-Muhtadi, J.; Solic, P. Energy meters evolution in smart grids: A review. *J. Clean. Prod.* **2019**, *217*, 702–715. [CrossRef]
11. Chen, C.K.; Lo, C.C. Optimal production run length for products sold with warranty in an imperfect production system with allowable shortages. *Math. Comput. Model.* **2006**, *44*, 319–331. [CrossRef]
12. Nižetić, S.; Djilali, N.; Papadopoulos, A.; Rodrigues, J.J.P.C. Smart technologies for promotion of energy efficiency, utilization of sustainable resources and waste management. *J. Clean. Prod.* **2019**, *231*, 565–591. [CrossRef]
13. Sarkar, B.; Moon, I. An EPQ model with inflation in an imperfect production system. *Appl. Math. Comput.* **2011**, *217*, 6159–6167. [CrossRef]
14. Sarkar, B. A production-inventory model with probabilistic deterioration in two-echelon supply chain management. *Appl. Math. Model.* **2013**, *37*, 3138–3151. [CrossRef]
15. Lund, H.; Østergaard, P.A.; Connolly, D.; Mathiesen, B.V. Smart energy and smart energy systems. *Energy* **2017**, *137*, 556–565. [CrossRef]

16. Edgar, T.F.; Pistikopoulos, E.N. Smart manufacturing and energy systems. *Comput. Chem. Eng.* **2018**, *114*, 130–144. [CrossRef]
17. Sana, S.; Goyal, S.K.; Chaudhuri, K.S. An imperfect production process in a volume flexible inventory model. *Int. J. Prod. Econ.* **2007**, *105*, 548–559. [CrossRef]
18. Louly, M.A.O.; Dolgui, A. Calculating safety stocks for assembly systems with random component procurement lead times: A branch and bound algorithm. *Eur. J. Oper. Res.* **2009**, *199*, 723–731. [CrossRef]
19. Jaber, M.Y.; Bonney, M.; Moualek, I. An economic order quantity model for an imperfect production process with entropy cost. *Int. J. Prod. Econ.* **2009**, *118*, 26–33. [CrossRef]
20. Sarkar, B.; Sana, S.S.; Chaudhuri, K. Optimal reliability, production lotsize and safety stock: An economic manufacturing quantity model. *Int. J. Manag. Sci. Eng. Manag.* **2010**, *5*, 192–202. [CrossRef]
21. Sana, S.S. An economic production lot size model in an imperfect production system. *Eur. J. Oper. Res.* **2010**, *201*, 158–170. [CrossRef]
22. Sarkar, B.; Ullah, M.; Kim, N. Environmental and economic assessment of closed-loop supply chain with remanufacturing and returnable transport items. *Comput. Ind. Eng.* **2017**, *111*, 148–163. [CrossRef]
23. Sarkar, M.; Hur, S.; Sarkar, B. Effects of variable production rate and time-dependent holding cost for complementary products in supply chain model. *Math. Probl. Eng.* **2017**, *2017*, 2825103. [CrossRef]
24. Sarkar, B.; Guchhait, R.; Sarkar, M.; Cárdenas-Barrón, L.E. How does an industry manage the optimum cash flow within a smart production system with the carbon footprint and carbon emission under logistics framework? *Int. J. Prod. Econ.* **2019**, *213*, 243–257. [CrossRef]
25. Cárdenas-Barrón, L.E.; Sarkar, B.; Garza, G.T. An improved solution to the replenishment policy for the EMQ model with rework and multiple shipments. *Appl. Math. Model.* **2013**, *37*, 5549–5554. [CrossRef]
26. Giotitsas, C.; Pazaitis, A.; Kostakis, V. A peer-to-peer approach to energy production. *Technol. Soc.* **2015**, *42*, 28–38. [CrossRef]
27. Tayyab, M.; Sarkar, B. Optimal batch quantity in a cleaner multi-stage lean production system with random defective rate. *J. Clean. Prod.* **2016**, *139*, 922–934. [CrossRef]
28. Omair, M.; Sarkar, B.; Cárdenas-Barrón, L.E. Minimum Quantity Lubrication and Carbon Footprint: A Step towards Sustainability. *Sustainability* **2017**, *9*, 714. [CrossRef]
29. Kim, M.S.; Sarkar, B. Multi-stage cleaner production process with quality improvement and lead time dependent ordering cost. *J. Clean. Prod.* **2017**, *144*, 572–590. [CrossRef]
30. Moon, I.; Shin, E.; Sarkar, B. Min–max distribution free continuous-review model with a service level constraint and variable lead time. *App. Math. Comput.* **2014**, *229*, 310–315. [CrossRef]
31. Sarkar, B.; Majumder, A.; Sarkar, M.; Dey, B.K.; Roy, G. Two-echelon supply chain model with manufacturing quality improvement and setup cost reduction. *J. Ind. Manag. Opt.* **2017**, *13*, 1085–1104. [CrossRef]
32. Kim, S.J.; Sarkar, B. Supply chain model with stochastic lead time, trade-credit financing, and transportation discounts. *Math. Probl. Eng.* **2017**, *2017*, 6465912. [CrossRef]
33. Majumder, A.; Guchhait, R.; Sarkar, B. Manufacturing quality improvement and setup cost reduction in an integrated vendor-buyer supply chain model. *Eur. J. Ind. Eng.* **2017**, *13*, 588–612. [CrossRef]
34. Soni, H.N.; Sarkar, B.; Mahapatra, A.S.; Majumder, S.K. Lost sales reduction and quality improvement with variable lead time and fuzzy costs in an imperfect production system. *RAIRO Oper. Res.* **2018**, *52*, 819–837. [CrossRef]
35. Sarkar, B.; Guchhait, G.; Sarkar, M.; Pareek, S.; Kim, N. Impact of safety factors and setup time reduction in a two-echelon supply chain management. *Robot. Comput. Int. Manuf.* **2019**, *55*, 250–258. [CrossRef]
36. Dey, B.K.; Sarkar, B.; Sarkar, M.; Pareek, S. An integrated inventory model involving discrete setup cost reduction, variable safety factor, selling-price dependent demand, and investment. *RAIRO Oper. Res.* **2019**, *53*, 39–57. [CrossRef]
37. Sarkar, B.; Saren, S.; Sinha, D.; Hur, S. Effect of unequal lot sizes, variable setup cost, and carbon emission cost in a supply chain model. *Math. Probl. Eng.* **2015**, *2015*, 469486. [CrossRef]
38. Biel, K.; Glock, C.H. On the use of waste heat in a two-stage production system with controllable production rates. *Int. J. Prod. Econ.* **2016**, *181*, 174–190. [CrossRef]
39. Kim, M.S.; Kim, J.S.; Sarkar, B.; Sarkar, M.; Iqbal, M.W. An improved way to calculate imperfect items during long-run production in an integrated inventory model with backorders. *J. Manuf. Syst.* **2018**, *47*, 153–167. [CrossRef]

40. Unver, U.; Kara, O. Energy efficiency by determining the production process with the lowest energy consumption in a steel forging facility. *J. Clean. Prod.* **2019**, *215*, 1362–1370. [CrossRef]

41. Kumar, L.R.; Yellapu, S.K.; Zhang, X.; Tyagi, R.D. Energy balance for biodiesel production processes using microbial oil and scum. *Bioresource Technol.* **2019**, *272*, 379–388. [CrossRef]

42. Morato, T.; Vaezi, M.; kumar, A. Assessment of energy production potential from agricultural residues in Bolivia. *Renew. Sustain. Energy Rev.* **2019**, *102*, 14–23. [CrossRef]

43. Ahmed, W.; Sarkar, B. Impact of carbon emissions in a sustainable supply chain management for a second generation biofuel. *J. Clean. Prod.* **2018**, *186*, 807–820. [CrossRef]

44. Darom, N.A.; Hishamuddin, H.; Ramli, R.; Nopiah, Z.M. An inventory model of supply chain disruption recovery with safety stock and carbon emission consideration. *J. Clean. Prod.* **2018**, *197*, 1011–1021. [CrossRef]

45. Assid, M.; Gharbi, A.; Hajji, A. Production and setup control policy for unreliable hybrid manufacturing-remanufacturing systems. *J. Manuf. Syst.* **2019**, *50*, 103–118. [CrossRef]

46. Sarkar, B. Mathematical and analytical approach for the management of defective items in a multi-stage production system. *J. Clean. Prod.* **2019**, *218*, 896–919. [CrossRef]

47. Caro-Ruiz, C.; Lombardi, P.; Richter, M.; Pelzer, A.; Mojica-Nava, E. Coordination of optimal sizing of energy storage systems and production buffer stocks in a net zero energy factory. *Appl. Energy* **2019**, *238*, 851–862. [CrossRef]

48. Dincer, I.; Ezzat, M.F. Solar Energy Production. *Compr. Energy Syst.* **2018**, *3*, 208–251.

49. Dincer, I.; Al-Zareer, M. Geothermal Energy Production. *Compr. Energy Syst.* **2018**, *3*, 252–303.

50. Dincer, I.; Rosen, M.A.; Al-Zareer, M. Electrochemical Energy Production. *Compr. Energy Syst.* **2018**, *3*, 416–469.

51. Dincer, I.; Rosen, M.A.; Al-Zareer, M. Chemical Energy Production. *Compr. Energy Syst.* **2018**, *3*, 470–520.

52. Dincer, I.; Bicer, Y. Photonic Energy Production. *Compr. Energy Syst.* **2018**, *3*, 707–754.

53. Lu, Y.; Peng, T.; Xu, X. Energy-efficient cyber-physical production network: Architecture and technologies. *Comput. Ind. Eng.* **2019**, *129*, 56–66. [CrossRef]

54. Kazemi, H.; Shokrgozar, M.; Kamkar, B.; Soltani, A. Analysis of cotton production by energy indicators in two different climatic regions. *J. Clean. Prod.* **2018**, *190*, 729–736. [CrossRef]

55. Kluczek, A. An energy-led sustainability assessment of production systems—An approach for improving energy efficiency performance. *Int. J. Prod. Econ.* **2019**, *216*, 190–203. [CrossRef]

56. Harris, T.M.; Devkota, J.P.; Khanna, V.; Eranki, P.L.; Landis, A.E. Logistic growth curve modeling of US energy production and consumption. *Renew. Sustain. Energy Rev.* **2018**, *96*, 46–57. [CrossRef]

57. Bruni, G.; Rizzello, C.; Santucci, A.; Alique, D.; Tosti, S. On the energy efficiency of hydrogen production processes via steam reforming using membrane reactors. *Int. J. Hydrog. Energy* **2019**, *44*, 988–999. [CrossRef]

58. Dehning, P.; Blume, S.; Dér, A.; Flick, D.; Herrmenn, C.; Thiede, S. Load profile analysis for reducing energy demands of production systems in non-production times. *Appl. Energy* **2019**, *237*, 117–130. [CrossRef]

59. Nordborg, M.; Berndes, G.; Dimitriou, I.; Henriksson, A.; Mola-Yudego, B.; Rosenqvist, H. Energy analysis of willow production for bioenergy in Sweden. *Renew. Sustain. Energy Rev.* **2018**, *93*, 473–482. [CrossRef]

60. Khalil, M.; Berawi, M.A.; Heryanto, R.; Rizalie, A. Waste to energy technology: The potential of sustainable biogas production from animal waste in Indonesia. *Renew. Sustain. Energy Rev.* **2019**, *105*, 323–331. [CrossRef]

61. Keen, S.; Ayres, R.U.; Standish, R. A note on the role of energy in oroduction. *Ecol. Econ.* **2019**, *157*, 40–46. [CrossRef]

62. Chen, X.; Li, C.; Tang, Y.; Li, L.; Li, L. Integrated optimization of cutting tool and cutting parameters in face milling for minimizing energy footprint and production time. *Energy* **2019**, *175*, 1021–1037. [CrossRef]

63. Liang, J.; Wang, Y.; Zhang, Z.H.; Sun, Y. Energy efficient production planning and scheduling problem with processing technology selection. *Comput. Ind. Eng.* **2019**, *132*, 260–270. [CrossRef]

Article

Optimum Design of a Transportation Scheme for Healthcare Supply Chain Management: The Effect of Energy Consumption

Jihed Jemai [1] and Biswajit Sarkar [2,*]

[1] Department of Industrial Engineering, Hanyang University, Seoul 04763, Korea
[2] Department of Industrial & Management Engineering, Hanyang University, Ansan, Gyeonggi-do 15588, Korea
* Correspondence: bsbiswajitsarkar@gmail.com

Received: 1 June 2019; Accepted: 16 July 2019; Published: 19 July 2019

Abstract: The perishability of blood platelets complicates the management of their supply chain. This paper studies the impact of energy consumption and carbon emissions of transportation activities in a blood platelet supply chain. Energy consumption and carbon emissions vary significantly, and the effective location-allocation of blood facilities is a key strategy for the optimal use of energy. The total cost of the supply chain for perishable products is minimized when energy consumption is optimized. The proposed model is too complex to be solved with existing methodologies; therefore, mathematical tools are used to solve it. A numerical experiment is carried out to validate the proposed model, and graphical representations are presented for better visualization of the study's outcomes. The results of the numerical studies confirm that the selected locations of blood facilities are optimal for the maximization of energy efficiency and minimization of the total cost.

Keywords: energy; healthcare supply chain management; platelets; perishability factor; location-allocation

---

1. Introduction

The operational management of any healthcare sector is becoming one of the most widely explored research fields. There has been an increasing demand for more systematic and efficient planning in order to meet the high standards of the healthcare sector. Platelets, a blood component with a high perishability rate, are considered to be a life-saving and valuable resource, and many patients' lives are dependent on their availability. Platelets are needed daily in almost every hospital as a life-saving measure for patients suffering from serious diseases and conditions, such as cancer, low platelet levels, bone marrow transplants, etc. Platelets can be collected from a donor by two different methods. The first is the traditional technique in which whole blood is drawn, and platelets must then be separated from other blood components within five hours of the donation. The second method is the "apheresis method". This technique not only saves time but also allows the other blood components to be returned to the donor's body immediately after isolating and collecting the platelets. Compared with the traditional technique, this method can be costly, but it is more efficient and can produce 6–10 times the number of platelets. Therefore, fewer donors are needed to produce the same number of platelets as the traditional method. Using the "apheresis method" can reduce the time interval between donations: donating whole blood is allowed every 56 days, while donors whose platelets are extracted using the "apheresis method" are eligible to donate platelets every seven days. In platelet supply chain management, the first step is to transfer donor platelets from blood facilities to a blood center. The latter not only tests, processes, stores, and distributes platelets to their demand points but also serves as a potential site for platelet donation. Blood platelets collected from different blood facilities must be transported to a blood center for the required tests before they are delivered to hospitals. Figure 1 shows the flow of the supply chain network of blood platelets.

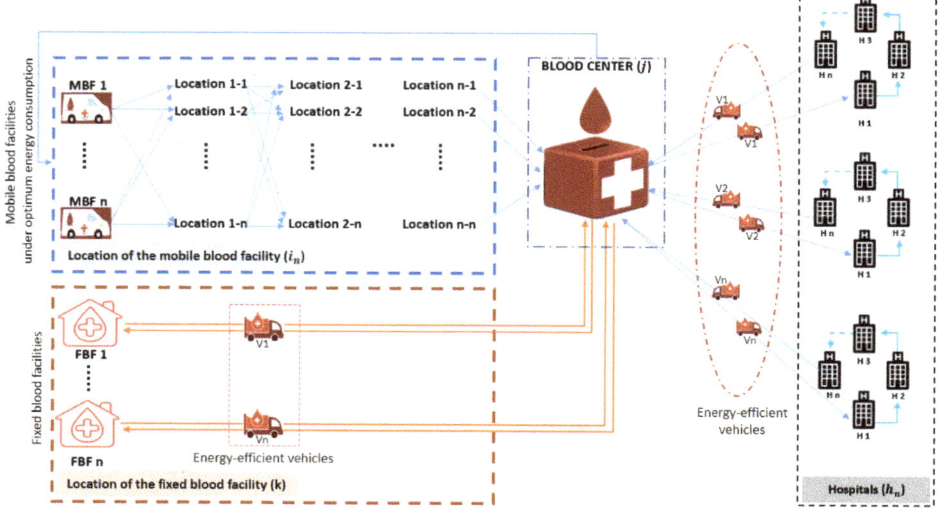

Figure 1. Platelet supply chain network with the effect of energy consumption.

A major challenge that complicates the supply chain management of platelets is their high perishability compared with other blood components. It has been proved that the lifespan of blood platelets ranges from five to seven days (Michael [1]); thus, their inventory must be managed efficiently. The limited shelf life of platelets is the reason that donors are always needed. The importance of platelet donation can be illustrated by the fact that one platelet donation can save the lives of three adults or 12 children, while 4–6 whole blood units can save only one patient (according to the NHSBT).

The objective of this paper is to develop a healthcare supply chain model that minimizes the total cost and energy consumption of the network by determining the optimal locations of mobile and fixed blood facilities to fulfill the demand of the demand points. The remainder of the paper is arranged as follows. Section 2 provides a brief literature review of the prior research pertinent to sustainable supply chain networks and problems encountered in the design of a blood supply chain network. Section 3 includes the problem definition and model formulation. In Section 4, a solution methodology is discussed, and numerical experiments are presented. Finally, conclusions and future research of the proposed model are given in Section 5.

2. Literature Review

Recently, the development of mathematical models to optimize the supply chain of blood platelets has drawn the attention of many researchers, and the number of studies in this field has increased significantly.

2.1. Healthcare Supply Chain Management

One of the main reviews with a focus on supply chain management in healthcare is the work of Nahmias [2]. This paper was the first to examine the relevant literature on the challenges faced when determining suitable ordering policies for both fixed-life perishable inventory and inventory characterized by continuous exponential decay. However, the review did not address the supply chain for blood and its components. Sha et al. [3] proposed an emergency blood supply scheduling model that primarily provides decision-making support for emergency blood supply without considering the effect of energy consumption. Jabbarzadeh et al. [4] specified a robust model of network design with a cost objective function of blood supply during and after disasters. Issues related to energy consumption were not studied in their paper. Using a real case study, Katsaliaki et al. [5] considered a game-based decision-making approach for delivering perishable products from donors to patients using the UK blood supply chain, but energy consumption and its related issues were not considered.

Duan and Liao [6] implemented a blood supply chain simulation-optimization framework with the objective of minimizing the outdate rate of blood, but they did not incorporate the effects of energy consumption. Arvan et al. [7] formulated a model for optimizing the blood supply chain network. They took several products and deterministic parameters into account and used the CPLEX solver to solve the model by using GAMS. Giannakis and Papadopoulos [8] addressed the operational perspective to achieve a sustainable supply chain by considering the risk management process, although energy-related concepts were not investigated. Osorio et al. [9] considered an integrated simulation-optimization model to take into account uncertain supply and demand, blood group rate, duration of validity constraints, and blood supply chain collection and production methods. Zahiri et al. [10] focused on the development of a healthcare supply chain network that accounts for blood group compatibility but excludes the effect of energy.

Ramezanian et al. [11] applied a mixed-integer linear model to the design of a blood supply chain network. They assumed that different social factors, such as distance, advertisement cost, and experience, affect the process of the decisions made by donors. The energy consumption factor was not included among the considered factors. Paydar et al. [12] provided a robust optimization model to design a bi-objective multi-period blood supply chain network for implementation during a disaster. They incorporated three echelons, namely, supply, processing, and delivery. The concept of energy consumption was not included in this study. Najafi et al. [13] looked at the uncertainty of blood demand and supply and the possibility of blood transshipment, but they did not consider the perishability factor or the effect of energy. Ensafian et al. [14] considered a platelet supply chain in which the demand is age-differentiated by the type of patient. Considering the apheresis method and the traditional method of platelet production, two mixed-integer programming models based on First In First Out and Last In First Out policies were developed.

Hosseinifard and Abbasi [15] focused on the design of a two-echelon blood supply chain with inventory centralization in the second echelon. They showed that centralizing hospital inventory is a key factor that can improve the sustainability and resilience of the blood supply chain, but the concept of energy consumption was neglected. Samani et al. [16] proposed a blood supply chain for disaster relief using a multi-objective mixed-integer linear programming model, which did not include the influence of energy consumption. Eskandari et al. [17] formulated a possible optimization model for a sustainable blood supply chain with multi-period and multi-objective data to address the uncertain conditions that exist during and after a disaster. The concept of energy consumption was not included in this study, and the perishability factor was not a part of the developed model. A two-stage stochastic programming problem for red blood cells was introduced by Hamdan and Diabat [18]. It simultaneously accounted for production, inventory, and location decisions, but the effect of energy consumption was not considered. Furthermore, healthcare supply chain management was not studied. The main optimization model of this paper is formulated on the basis of the mathematical formulation of Eskandari et al. [17], who designed a possibilistic model for multi-objective and multi-period sustainable blood supply chain in the presence of uncertainty due to unforeseeable conditions during and after a disaster.

2.2. Sustainable Energy-Efficient Supply Chain Management (SEESCM)

Beamon [19] studied the opportunities and challenges, such as energy availability and environmental protection, that might face the supply chain in the future, as well as their effects on the management, design, and integration of the supply chain. The author studied the supply chain from a general perspective without specifying the type of supply chain. Rentizelas et al. [20]

addressed the logistics of biomass, which is a renewable energy source, and the storage problem in a multi-biomass supply chain. In this paper, the authors focused only on the biomass supply chain. Halldórsson and Kovács [21] designed logistics solutions to the issue of climate change to determine a sustainable supply chain while accounting for energy efficiency. However, the supply chain of a perishable product for healthcare was not discussed in this paper. Gold and Seuring [22] conducted a literature review of logistics issues and supply chains in bioenergy production. Healthcare supply chain management was not considered in this review. Although Pan et al. [23] reviewed some implementation strategies of a waste-to-energy (WTE) supply chain in a circular economic system, they did not study the healthcare supply chain management of perishable products.

Luthra et al. [24] used Analytical Hierarchy Process to evaluate the obstacles that may be faced when adopting sustainable production and energy consumption in a supply chain. However, they did not include healthcare supply chain management in this study. Hong et al. [25] conducted a multi-regional structural analysis of the consumption of energy in a construction supply chain. They focused only on the impact of energy consumption in the construction industry in China and did not consider any healthcare issues. Marzband et al. [26] evaluated energy management systems in real-time for smart hybrid home microgrids but did not assess the supply chain of perishable products in the healthcare sector. Aziziankohan et al. [27] investigated the impact of queuing theory on the reduction of waiting times, the optimization of energy consumption in a green supply chain, and the decrease in pollution. Issues related to healthcare were not explored. Centobelli et al. [28] conducted a review of energy efficiency and environmental sustainability in the context of supply chain management to identify research gaps and research trends; however, healthcare supply chain management was not mentioned.

Singh et al. [29] investigated the performance of a food supply chain and its dependence on the performance of the participants to achieve an energy-efficient and sustainable supply chain; the topic of healthcare supply chain management was not studied. Sarkar et al. [30] evaluated the effect of energy and the failure rate in a multi-product smart production system. However, healthcare supply chain management of perishable products was not explored. Sarkar et al. [31] worked on developing a multi-objective optimization model. Their objectives were to minimize total costs, carbon emissions, and the cost of energy consumption, although they did not include issues related to healthcare supply chain management. Marchi et al. [32] formulated a two-stage model while considering the opportunity to invest in new energy-efficient solutions with better performance and better financial conditions, and members of the supply chain exerted an important influence. Healthcare supply chain management was not a part of this study. Fontes et al. [33] identified the latest dynamic system contributions and trends in a renewable energy supply chain without studying the healthcare supply chain of perishable products.

2.3. Green Supply Chain Management

Zhu et al. [34] presented a study on green supply chain management in the automobile industry of China but did not cover the healthcare sector. Srivastava [35] reviewed the literature on green supply chain management; however, the healthcare sector was not addressed, and the effect of energy was not introduced. Yeh and Chuang [36] developed an optimum model for the selection of green partners. The study had four metrics: cost, time, product quality, and appraisal score. The effect of energy was not considered in this study, and the healthcare sector was not examined. Hua et al. [37] investigated possible methods for managing carbon footprints in inventory with a carbon emissions technique. Kumar et al. [38] focused on green supply chain practices that could be adopted by electrical and electronics manufacturing industries. The authors specified the field of application of the study. The optimum consumption of energy was not considered, and the healthcare sector was not studied.

Pan et al. [39] provided an optimization model to reduce greenhouse gas emissions from freight transport. Neither the effect of energy nor healthcare supply chain management were applied. Bazan et al. [40] introduced two models of greenhouse gas emissions from production and transportation activities. They included energy consumption and examined different decisions but did not study any healthcare-related issues. Ahmed et al. [41] focused on the effect of carbon emissions in a sustainable second-generation supply chain of biofuel with a cost minimization objective. However, the effect of energy consumption was not included, nor was the healthcare supply chain management of perishable product studied.

2.4. Traditional Supply Chain Management with a Single Objective

Sarkar [42] developed a model of product inventory for a deteriorating product in a two-echelon supply chain in which he assumed only probabilistic deterioration with multiple distributions, but he did not consider healthcare or energy issues. In another study, Sarkar et al. [43] developed a mathematical model with the aim of reducing the total cost of the supply chain while considering the variable costs of transportation and carbon emissions due to multiple shipments. Energy consumption was not considered in the model, and healthcare supply chain management of perishable product was not studied. Habib and Sarkar [44] designed a location-allocation model composed of a two-phase framework for sustainable debris management during disasters given an uncertain environment. The concepts of healthcare and energy consumption were not considered in the study. Sarkar et al. [45] developed an integrated inventory model that was based on variable transportation costs to maintain a single-setup multi-delivery policy for the reduction of transportation costs. However, healthcare and energy consumption were not investigated.

Feng et al. [46] studied an integrated inventory model consisting of a single supplier and several buyers with shared warehouse capacity via transshipment. They did not consider healthcare or energy issues in this study. Sarkar et al. [47] analyzed multi-level delay-in-payment, variable transportation costs, and variable carbon emission costs in order to minimize the total costs of a sustainable supply chain under a policy of single-setup multi-delivery without mentioning specific products. The effects of energy consumption and healthcare supply chain management were not included in the study. Shi et al. [48] incorporated the development of a supply chain with demand uncertainty in different power structures to investigate the effect of sustainable investment on profit and emissions. Energy and healthcare issues were not addressed.

Iqbal et al. [49] analyzed two types of supply chains. The objective was to minimize total cost per unit time in the network. In the first supply chain, the products of both systems were deteriorated; in the second one, the products of the first supply chain were deteriorated. Neither the impact of energy consumption nor healthcare issues were studied. Mishra et al. [50] provided a mathematical model for a sustainable electricity supply chain with the assumption of linear price-dependent customer demands. In the proposed model, price was a decision variable for carbon emissions and setup costs, but energy consumption was not a studied factor. They did not consider healthcare supply chain management in their research.

From reviewing the literature, a research gap is apparent. The following main research contributions are presented to close some of these gaps in the study of blood supply chain network design. The presented paper contributes in two different ways. First, the proposed model includes various characteristics that were not considered in previous research on blood platelet supply chain management. This paper presents the design of a comprehensive blood supply chain model in which (1) the limited shelf life of blood platelets is considered, (2) the distance between nodes in the network is determined in order to maintain efficient energy consumption and reduce carbon emissions during transportation activities, and (3) decision-making related to the determination of the optimal facility location-allocation is included, with the aim of minimizing the total cost of the supply chain. Furthermore, the presented model includes other aspects that have been rarely examined by studies on the blood supply chain; these topics include methods of collecting blood platelets in mobile blood facilities and the blood diffusion process between the blood center and demand points, with multi-delivery mapping a considered factor.

- Not Applicable (NA) means that the concept was not studied in the specified paper;
- Applicable means that the concept was covered by the specified paper.

Table 1 illustrates the novelty of the presented research. The listed studies were found by using the following keywords: supply chain management and optimum energy, blood collection, platelets, location, and perishability. Each research paper presented in the above literature review was studied while considering these keywords. The investigation of each research paper aimed to verify whether each concept was applicable or not applicable to identify gaps in the literature. To the best of the authors' knowledge, no research has combined the five concepts in one paper.

Table 1. Contributions of different authors.

Authors	SCM and the Optimum Energy (OE)	Blood Collection	Platelets	Location	Perishability
Nahmias [2]	SCM	NA	NA	NA	Applicable
Sha et al. [3]	SCM	NA	NA	NA	NA
Sarkar et al. [42]	SCM	NA	NA	NA	Applicable
Jabbarzadeh et al. [4]	SCM	Applicable	NA	NA	NA
Katsaliaki et al. [5]	SCM	Applicable	NA	NA	Applicable
Duan and Liao [6]	SCM	Applicable	NA	NA	NA
Bazan et al. [40]	SCM	NA	NA	NA	NA
Arvan et al. [7]	SCM	Applicable	NA	NA	NA
Hong et al. [25]	SCM	NA	NA	NA	NA
Habib and Sarkar [44]	SCM	NA	NA	Allocated	NA
Zahiri et al. [10]	SCM	Applicable	NA	Allocated	Applicable
Ramezanian et al. [11]	SCM	Applicable	NA	NA	NA
Paydar et al. [12]	SCM	Applicable	NA	NA	Applicable
Osorio et al. [9]	SCM	Applicable	NA	Allocated	Applicable
Najafi et al. [13]	SCM	Applicable	NA	NA	NA
Ensafian et al. [14]	SCM	Applicable	Applicable	NA	Applicable
Ahmed et al. [41]	SCM	NA	NA	Allocated	NA
Sarkar et al. [30]	SCM	NA	NA	NA	NA
Hosseinifard and Abbasi [15]	SCM	Applicable	NA	NA	NA
Singh et al. [29]	SCM	NA	NA	NA	NA
Samani et al. [16]	SCM	Applicable	NA	NA	NA
Eskandari et al. [17]	SCM	Applicable	Applicable	NA	NA
Hamdan and Diabat [18]	SCM	Applicable	NA	Allocated	NA
Mishra et al. [50]	SCM	NA	NA	NA	NA
Iqbal et al. [49]	SCM	NA	NA	NA	Applicable
Proposed model	SCM & OE	Applicable	Applicable	Allocated	Applicable

3. Problem Definition, Notation, and Assumptions

This section defines the problem and provides the notation and assumptions of the proposed model.

3.1. Problem Definition

The supply chain under study consists of fixed blood facilities, mobile blood facilities, blood centers, and demand points (hospitals). Depending on the working hours of the employees and the daily demand for platelets, the location of mobile blood facilities varies from one period to another. Platelets can be donated at a mobile blood facility (MBF), a fixed blood facility (FBF), or even at a blood center. The platelets collected in mobile blood facilities are delivered by the mobile facility itself to the designated blood center in the region once the daily demand is met or the employees' working hours end. Efficient location-allocation of mobile blood facilities that considers the distance between candidate locations and the blood center can have a significant positive effect on energy consumption and can thus lead to cost reduction. On the other hand, the location of fixed blood facilities is an important decision that must be made carefully in order to guarantee the sustainability of the healthcare supply chain and ensure that transportation activities between fixed blood facilities and the blood center are carried out with consideration given to energy consumption. Blood platelets collected in a fixed blood facility are transported to the assigned blood center by specific vehicles with optimal energy consumption and known capacities. These vehicles must return to their starting point after completing their mission to prepare for the next mission. The platelets received by the blood center are subjected to several tests for a maximum of two days to verify that the blood is free from diseases. The blood is then classified according to its age, followed by storage and preparation for distribution to demand points. Blood centers then have to meet the demands of each hospital by assigning one energy-efficient vehicle to each route to visit one or more demand points in each trip during the planning period t. By the end of the planning period t, these vehicles must return to their starting points (the blood center). A method used to draw platelets is the "apheresis method", which is able to produce more platelets than the traditional method; thus, fewer donors are needed to produce the daily demand for platelets.

The proposed model for platelet supply chain management that incorporates the effect of energy, therefore, aims to minimize the overall cost of the supply chain. The notation in the model's formulation consists of indices, decision variables, and parameters and is given in Appendix A. The assumptions of this problem are specified in Section 3.3. The flow diagram of healthcare supply chain management with optimal consumption of energy is given in Figure 2.

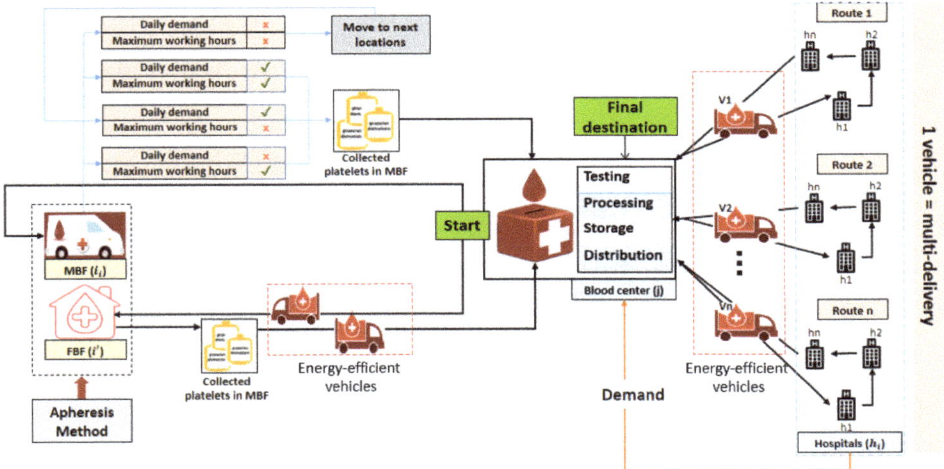

Figure 2. Flow diagram of healthcare supply chain management with optimal energy consumption.

3.2. Notation

The notation used in this paper is provided in Appendix A.

3.3. Assumptions

To structure the proposed model, the following assumptions were made to specify boundaries in the formulation of the model.

- The effect of energy is considered with optimal energy consumption in a healthcare supply chain [51] for blood platelets.
- Each hospital H can be served by one energy-efficient vehicle V, and each vehicle has to return to the blood center at location j after completing its mission [52]. Each vehicle V is assigned to one or more hospitals [53]. The transportation of platelets must maintain the optimal energy level [54].

- The demand $Dm_{h_i t}$ that occurs in one working shift must be fully covered within the same shift in one trip to maintain energy efficiency [55]. In addition, the maximum demand of one hospital per day is assumed to be 30 units or less [56].
- The speed of each vehicle is considered to be fixed [57]. Thus, energy consumption is maintained, and each vehicle starts and ends its trip at the starting point [58]. The distances between the nodes of the supply chain are known, and the number, locations, and capacities of the MBF, FBF, blood center, and delivery vehicles for optimal energy consumption are known [59].
- Only blood platelets are considered in this study, and their lifetime L is known and limited [60]. The age a of the platelet units is known (less than three days: young; more than three days: old) [61]. Only a mobile blood facility is able to move from one site to another in the next planning period t, and its movement to another site depends on the working hours of the employees and whether the required quantity is achieved [62].
- The transportation costs, including carbon emissions and energy consumption [63], are proportional to the distance traveled.
- Blood wastage refers to the rate of lost blood α during the testing process. Thus, wastage costs are incurred. The expired quantity of blood (Qu_t and Qp_t in blood centers and fixed blood facilities, respectively) imposes extra costs. It is assumed that the volume of 1 unit of blood platelets is fixed and equal to 180 mL.

4. Model Formulation

The main objective of the proposed mathematical model is to minimize the total cost of the blood platelet supply chain by incorporating the effect of the energy expended by delivery activities between network nodes. Minimizing the total cost of a supply chain is always a central focus. The main objective of the proposed model is an economic objective in which the total cost of the supply chain is minimized, and the optimal locations of mobile blood facilities and fixed blood facilities are determined.

4.1. Transportation Costs Due to the Effect of Energy

Minimizing transportation costs can lower the total cost of the supply chain by a considerable amount. Transportation costs are considered to have a significant role in supply chain management. The overall transportation cost determines the costs related to the transportation activities between nodes in the network.

Transportation costs from a fixed blood facility to a blood center with optimal energy consumption

The fixed transportation costs due to the effect of energy consumption per km with the optimal consumption of energy is multiplied by the distance between the fixed blood facility and the blood center $Tc\, d_{kj}$ to determine the total fixed costs. The total carbon emission costs are determined by multiplying the cost of CO_2 per km by the total distance between the fixed blood facility and the blood center $CE_c\, d_{kj}$. Thus, the following expression is determined:

$$Trc_{FBF}^{BC} = \sum_{k=1}^{K}\sum_{j=1}^{J}\sum_{t=1}^{T}\left[(Tc + CE_c)\, d_{kj}\right] U_{vc_v t} \tag{1}$$

Transportation costs from a mobile blood facility to a blood center with efficient energy consumption

To determine the transportation costs required during the delivery of blood platelets collected from the location of a mobile blood facility to the blood center, the fixed transportation costs due to the effect of energy per km are multiplied by the total distance between two nodes in the network $(Tc\, d_{i_n jt})$. The carbon emission cost is obtained by multiplying the total distance by the cost of CO_2 per km, i.e., $(CE_c\, d_{i_n jt})$. Thus, the below expression is obtained:

$$Trc_{MBF}^{BC} = \sum_{i_n=1}^{I} \sum_{j=1}^{J} \sum_{t=1}^{T} [(Tc + CE_c)\, d_{i_n jt}]\, M_{cm\,i_n jt} \tag{2}$$

Transportation costs due to energy efficiency from a blood center to hospital

The fixed transportation costs per km with optimal consumption of energy are multiplied by the total distance between the blood center and the hospitals ($Tc\, d_{jh_i t}$), and the total carbon emission costs are determined by multiplying the carbon emission cost per km and the total distance covered during the delivery process ($CE_c\, d_{jh_i t}$). By the implied summation, the total transportation costs are calculated by the following expression:

$$Trc_{H}^{BC} = \sum_{j=1}^{J} \sum_{h_i=1}^{H} \sum_{t=1}^{T} [(Tc + CE_c)\, d_{jh_i t}]\, U_{vc_v t} \tag{3}$$

Moving costs of a mobile blood facility

A mobile blood facility can change its location in each planning period. To determine the moving costs of a mobile blood facility, the fixed transportation costs per km with optimal energy consumption are multiplied by the total distance that a mobile blood facility covers when moving from one location to another during planning periods ($Tc\, d_{i_{n-1}i_n}$). The carbon emitted by the mobile blood facility is calculated by multiplying the carbon emission cost per km by the distance ($CE_c\, d_{i_{n-1}i_n}$); thus, the following summation is formulated:

$$Trc_{MBF} = \sum_{i_n=1}^{I} \sum_{t=1}^{T} [(Tc + CE_c)\, d_{i_{n-1}i_n}]\, G_{i_{n-1}i_n t} \tag{4}$$

4.2. Inventory Holding Costs

Inventory holding costs are the costs associated with storing the blood platelets that remain undistributed. These costs represent one component of the total inventory costs, along with shortage costs and ordering costs.

Inventory holding costs at a blood center

Holding costs are incurred to keep platelets in the blood center's inventory. The holding costs of blood units in the blood center are based on the inventory level. Thus, the unit holding cost at a blood center is multiplied by the inventory level of blood platelets in the blood center. Consequently, the inventory holding costs at the blood center are calculated by the expression below:

$$IHC_{BC} = \sum_{j=1}^{J} \sum_{a=1}^{A} \sum_{t=1}^{T} hc_{ajt}\, IL_{ajt} \tag{5}$$

Inventory holding costs at a fixed blood facility

Similar to inventory holding costs at the blood center, fixed blood facilities have inventory costs that are obtained by the following expression:

$$IHC_{FBF} = \sum_{k=1}^{K} \sum_{a=1}^{A} \sum_{t=1}^{T} hc_{akt}\, IL_{akt} \tag{6}$$

4.3. Shortage Costs

The shortage cost is calculated by multiplying the cost of one unit of shortage in blood with age a by the quantity of the shortage in blood during the planning period t. Thus, the following expression is obtained.

$$Shc = \sum_{a=1}^{A} \sum_{j=1}^{J} \sum_{t=1}^{T} \pi \, Qs_{ajt} \tag{7}$$

4.4. Wastage Costs

The wastage cost represents the quantity of blood consumed during the testing process. This cost is calculated by first summing the quantity of the blood delivered from the blood facilities to the blood center and the quantity of blood platelets collected at the blood center $(Q_{i_njt} + Q_{kjt} + Q_{jt})$; then, this sum is multiplied by the production costs for one unit of blood platelets at the blood center. Thus, the following expression is obtained:

$$Wc = \sum_{j=1}^{J} \sum_{t=1}^{T} (Q_{i_njt} + Q_{kjt} + Q_{jt}) \, \alpha \, pc_{jt} \tag{8}$$

4.5. Perishable Blood Costs

Because blood platelets are perishable products, after a certain time, they cannot be used to save lives and will be deteriorated. Thus, the costs related to outdated blood platelets are regarded as perishability costs. Blood platelets can be stored in the inventory of either the blood center or the fixed blood facility. Therefore, perishability costs apply to both locations.

Perishable blood costs at a blood center

The costs of expired blood are obtained by multiplying the cost of one expired blood unit Cu by the total quantity of expired blood units $\beta \, IL_{ajt}$, where β is the perishability rate at the blood center. Thus, the following expression is obtained:

$$Pc_{BC} = \sum_{j=1}^{J} \sum_{a=1}^{A} \sum_{t=1}^{T} Cu \, \beta \, IL_{ajt} \tag{9}$$

Perishable blood costs at a fixed blood facility

The costs of outdated blood are obtained by multiplying the cost of one expired blood unit with the total quantity of expired blood units $\lambda \, IL_{akt}$, where λ is the perishability rate at a fixed blood facility. Thus, the following equation is used:

$$Pc_{FBF} = \sum_{k=1}^{K} \sum_{a=1}^{A} \sum_{t=1}^{T} Uc \, \lambda \, IL_{akt} \tag{10}$$

4.6. Operational Costs

Operating costs are the expenses related to the operation of each blood facility and blood center in order to produce and collect blood platelets. Thus, operational costs at mobile blood facilities, fixed blood facilities, and blood centers are calculated using the expressions defined below.

Operational costs at a mobile blood facility

Blood platelets are drawn using the apheresis method, so the costs of producing platelets at a mobile blood facility are obtained by multiplying the quantity collected Q_{i_njt} by the unit operational cost at the mobile blood facility in one period. Thus, the following expression is obtained:

$$Oc_{MBF} = \sum_{t=1}^{T} \sum_{i_n=1}^{I} \sum_{j=1}^{J} APHc_{i_n t} \, Q_{i_n jt} \tag{11}$$

Operational costs at a fixed blood facility

The operational costs of producing blood platelets through the apheresis method at a fixed blood facility are obtained by the below expression, in which Q_{kjt} is the quantity of platelets collected using the apheresis method at the fixed blood facility during the planning period t, and $APHc_{kt}$ is the unit operational cost at the fixed blood facility in period t.

$$Oc_{FBF} = \sum_{t=1}^{T} \sum_{k=1}^{K} \sum_{j=1}^{J} APHc_{kt} \, Q_{kt} \tag{12}$$

Operational costs at a blood center

Operational costs refer to the costs of producing platelets through the apheresis method at a blood center, and they are calculated by multiplying the quantity of platelets by the unit operational cost at the blood center. The operational costs at the blood center are calculated by the expression below:

$$Oc_{BC} = \sum_{j=1}^{J} \sum_{t=1}^{T} APHc_{jt} \, Q_{jt} \tag{13}$$

Production costs at a blood center

The collected blood platelets must undergo several tests in the blood center for a period of two days to confirm that the platelets are disease-free before being transferred to the patient's body. Thus, the production costs are calculated by multiplying the unit operational cost at the blood center by the quantity of blood platelets collected from donors at mobile and permanent blood facilities and the blood center in planning period t, where $pq_{jt} = (Q_{i_n jt} + Q_{kjt} + Q_{jt})(1 - \alpha)$. Thus, the below expression is applied:

$$PrC_{BC} = \sum_{j=1}^{J} \sum_{t=1}^{T} pc_{jt} \, pq_{jt} \tag{14}$$

4.7. Establishment Costs

Establishment costs or setup costs are the costs related to establishing fixed blood facilities at candidate locations and setting up mobile blood facilities to collect blood platelets from different sites. These costs also include the setup costs of the delivery vehicles needed to distribute the blood platelets between nodes in the supply chain.

Fixed costs of opening a fixed blood facility

The costs of establishing a fixed blood facility are obtained by multiplying the costs of establishment EF_{kc_f} by the binary variable W_{kc_f}, which is equal to 1 if a fixed blood facility is opened at location k and 0 otherwise. The summation of the establishment costs is represented below.

$$EC_{FBF} = \sum_{k=1}^{K} \sum_{c_f=1}^{C} EF_{kc_f} \, W_{kc_f} \tag{15}$$

Fixed costs of setting up a mobile blood facility

The costs of establishing a mobile blood facility are obtained by multiplying the costs of establishment $EM_{i_n c_m}$ by the binary variable $X_{i_n c_m}$, which is equal to 1 if a mobile blood facility is opened at location i_n and 0 otherwise. The summation of the establishment costs is represented below.

$$EC_{MBF} = \sum_{i_n=1}^{I} \sum_{c_m=1}^{C} EM_{i_n c_m} X_{i_n c_m} \tag{16}$$

Fixed costs of setting up delivery vehicles

To deliver the blood, vehicles with specific equipment must be available, and the costs of setting up such vehicles are calculated as shown in the expression below, i.e., by multiplying the costs of setting up the delivery vehicles by a binary variable that is equal to 1 if a vehicle with capacity c_v is set up and 0 otherwise.

$$SC_V = \sum_{c_v=1}^{C} \sum_{t=1}^{T} VC_{c_v} U_{vc_v t} \tag{17}$$

4.8. Constraints

The constraints of the designed model are presented in this subsection.

The first constraint is that only one blood center of size c_b is open at location j. Y_{jc_b} is a binary variable that is equal to 1 if a blood center with capacity c_b is established at site j.

$$\sum_{c_b=1}^{C} Y_{jc_b} \leq 1, \ \forall j \tag{18}$$

Constraint (19) states that one fixed blood facility can be established at location k with capacity c_f.

$$\sum_{c_f=1}^{C} W_{kc_f} \leq 1, \ \forall k \tag{19}$$

Constraints (20) and (21) stipulate that each blood facility, whether mobile or fixed, is assigned to one blood center.

$$\sum_{i_n=1}^{I} M_{c_m i_n jt} \leq 1, \ \forall j, t \tag{20}$$

$$\sum_{j=1}^{J} F_{kjt} \leq 1, \ \forall k, t \tag{21}$$

Constraint (22) indicates that only one mobile blood facility can move from location i_{n-1} to location i_n in each period t.

$$\sum_{i_n=1}^{I} G_{i_{n-1} i_n t} \leq 1, \ \forall t \tag{22}$$

Constraint (23) ensures that a mobile facility cannot move to a location where a mobile blood facility has already been located.

$$\sum_{i_n=1}^{I} G_{i_{n-1} i_n t} \leq \sum_{i_n=1}^{I} G_{i_{n-1} i_n, t-1}, \ \forall t \tag{23}$$

Constraints (24) and (25) state that the collected blood from each mobile and fixed blood facility should be appropriate given their maximum capacities c_m and c_f, respectively.

$$\sum_{i_n=1}^{I} Q_{i_n jt} \leq c_m \, X_{i_n c_m}, \ \forall \, j, t \tag{24}$$

$$\sum_{k=1}^{K} Q_{kjt} \leq c_f \, W_{kc_f}, \ \forall \, j, t \tag{25}$$

Constraint (26) specifies that the blood collected in each mobile and fixed blood facility for delivery to the blood center should not exceed the maximum capacity c_b of the blood center.

$$\sum_{i_n=1}^{I} Q_{i_n jt} + \sum_{k=1}^{K} Q_{kjt} \leq c_b \, Y_{jc_b}, \ \forall \, j, t \tag{26}$$

Constraint (27) indicates that the amount of wasted blood should be included when determining the amount of blood produced in the blood center at site j.

$$\sum_{j=1}^{J} pq_{jt} = \left(\sum_{i_n=1}^{I} \sum_{j=1}^{J} Q_{i_n jt} + \sum_{k=1}^{K} \sum_{j=1}^{J} Q_{kjt} + \sum_{j=1}^{J} Q_{jt} \right) (1 - \alpha), \ \forall \, t \tag{27}$$

Constraint (28) ensures that hospitals can be allocated to blood centers only if a blood center is open.

$$H_{h_i jt} \leq Y_{jc_b}, \ \forall \, j, t \tag{28}$$

Constraint (29) stipulates that each hospital can be assigned to only one blood center.

$$\sum_{h_i=1}^{H} \sum_{j=1}^{J} H_{h_i jt} \leq 1, \ \forall \, t \tag{29}$$

Constraint (30) shows that shortage is allowed but should not exceed the maximum shortage permitted for blood with age a in blood center j during planning period t.

$$Qs_{ajt} \leq Qs_{ajt}^{max} \ \forall \, a, j, t \tag{30}$$

Constraint (31) addresses the number of blood centers that can be assigned to hospitals; specifically, one blood center can be assigned to more than one hospital.

$$\sum_{j=1}^{J} B_{jh_i t} \geq 1, \ \forall \, h_i, t \tag{31}$$

Constraints (32) and (33) state that no blood facility can be assigned to a closed blood center.

$$M_{c_m i_n jt} \leq Y_{jc_b}, \ \forall \, i_n, j, c_b, t \tag{32}$$

$$F_{kjt} \leq Y_{jc_b}, \ \forall \, k, j, c_b, t \tag{33}$$

Constraint (34) indicates that the total produced quantity of blood platelets in period t must fulfill the total demand in period t.

$$\sum_{h_i=1}^{H} Dm_{h_i t} \leq \sum_{j=1}^{J} pq_{jt}, \ \forall \, t \tag{34}$$

Constraint (35) specifies that the demand of hospital h_i must be fulfilled by a single vehicle.

$$\sum_{v=1}^{V} y_{h_i v} = 1 \; \forall \, h_i \tag{35}$$

Constraint (36) ensures that the demand of the blood center can cover the demand of all the demand points in time period t.

$$\sum_{j=1}^{J} Dm_{jt} \geq \sum_{h_i=1}^{H} Dm_{h_i t} \; \forall \, t \tag{36}$$

Constraint (37) states that the blood platelets collected from all facilities can cover the demand of the blood center in time period t.

$$\sum_{i_n=1}^{I} \sum_{j=1}^{J} Q_{i_n jt} + \sum_{k=1}^{K} \sum_{j=1}^{J} Q_{kjt} \geq \sum_{j=1}^{J} Dm_{jt} \; \forall \, t \tag{37}$$

Constraints (38)–(40) demonstrate the relation between the inventory level, the perishability rate, the wastage rate, and the total quantity of collected blood platelets.

$$Qp_t = \lambda \, IL_{akt} \tag{38}$$

$$Qu_t = \beta \, IL_{ajt} \tag{39}$$

$$Qw_t = \alpha \, (Q_{i_n jt} + Q_{kjt} + Q_{jt}) \tag{40}$$

Constraint (41) is a maximization function that restricts the costs of outdated platelets to be equal to the number of platelet units in the inventory at L minus the amount of consumption in that period. The FIFO policy is applied to minimize the perishability rate. This constraint is valid for every period beyond the lifespan of the unit.

$$Qu_t = max\{0, IL_{aj,t-L} - \sum_{h_i=1}^{H} Q_{jh_i t} \, Qu_{t-L}\} \; \forall \, h_i \in H, t \in T : t \geq L+1, a \in A \tag{41}$$

Constraints (42) and (43) calculate the blood platelet inventory level constraints at the fixed blood facilities and blood centers, respectively.

$$IL_{ajt} = IL_{aj,t-1} - Qw_t - Qu_t + pq_{jt} - \sum_{j=1}^{J} \sum_{h_i=1}^{H} \sum_{t=1}^{T} Q_{jh_i t} \; \forall \, j \in J, t \in T, a \in A \tag{42}$$

$$IL_{akt} = IL_{ak,t-1} - Qp_t + Q_{kt} - \sum_{k=1}^{K} \sum_{j=1}^{J} \sum_{t=1}^{T} Q_{kjt} \; \forall \, k \in K, t \in T, a \in A \tag{43}$$

Constraints (44) and (45) specify that the inventory level of a blood center and fixed blood facilities should not exceed their maximum capacities c_b and c_f, respectively.

$$IL_{ajt} \leq c_b, \; \forall j, t \tag{44}$$

$$IL_{akt} \leq c_f, \; \forall k, t \tag{45}$$

Constraints (46) and (47) represent the binary and non-negativity restrictions on the decision variables.

$$W_{kc_f}, X_{i_n c_m}, Y_{jc_b}, M_{c_m i_n jt}, F_{kjt}, H_{h_i jt}, G_{i_{n-1} i_n t}, U_{vc_v t} \in 0, 1 \forall i_n \in I, k \in K, c_f, c_m, c_b, c_v \in C, j \in J, \\ t \in T, h_i \in H \tag{46}$$

$$Q_{kjt}, Q_{i_njt}, Q_{kt}, Q_{jt}, U_{c_mt}, Q_{jh_it}, IL_{ajt}, IL_{akt}, Qs_{ajt}, Qs_{ajt}^{max}, Qw_t, Qu_t, pq_{jt} \geq 0 \,\forall\, i_n \in I, k \in K,$$
$$j \in J, t \in T, c_v, c_m, c_b, c_f \in C, v \in V, a \in A, h_i \in H \tag{47}$$

4.9. Objective Function

The mathematical model was developed with the objective of reducing the total cost of healthcare supply chain management by optimizing the energy consumption, which includes the minimization of transportation costs, holding costs, wastage costs, expired blood costs, and establishment costs of blood facilities and blood centers. Thus, the following expression is obtained.

$$
\begin{aligned}
Min\ Z1 = &\sum_{k=1}^{K}\sum_{j=1}^{J}\sum_{t=1}^{T}\left[(Tc + CE_c)\, d_{kj}\right] U_{vc_vt} + \\
&\sum_{i_n=1}^{I}\sum_{j=1}^{J}\sum_{t=1}^{T}\left[(Tc + CE_c)\, d_{i_njt}\right] M_{cmi_njt} + \\
&\sum_{j=1}^{J}\sum_{h_i=1}^{H}\sum_{t=1}^{T}\left[(Tc + CE_c)\, d_{jh_it}\right] U_{vc_vt} + \sum_{i_n=1}^{I}\sum_{t=1}^{T}\left[(Tc + CE_c)\, d_{i_{n-1}i_n}\right] G_{i_{n-1}i_nt} + \\
&\sum_{j=1}^{J}\sum_{a=1}^{A}\sum_{t=1}^{T}hc_{ajt}\, IL_{ajt} + \sum_{j=1}^{J}\sum_{a=1}^{A}\sum_{t=1}^{T}hc_{akt}\, IL_{akt} + \sum_{a=1}^{A}\sum_{j=1}^{J}\sum_{t=1}^{T}\pi\, Qs_{ajt} + \\
&\sum_{j=1}^{J}\sum_{t=1}^{T}(Q_{i_njt} + Q_{kjt} + Q_{jt})\,\alpha\, pc_{jt} + \sum_{j=1}^{J}\sum_{a=1}^{A}\sum_{t=1}^{T}Cu\,\beta\, IL_{ajt} + \sum_{k=1}^{K}\sum_{a=1}^{A}\sum_{t=1}^{T}Uc\,\lambda\, IL_{akt} + \\
&\sum_{t=1}^{T}\sum_{i_n=1}^{I}\sum_{j=1}^{J}APHc_{i_nt}\, Q_{i_njt} + \sum_{t=1}^{T}\sum_{k=1}^{K}APHc_{kt}\, Q_{kt} + \sum_{j=1}^{J}\sum_{t=1}^{T}APHc_{jt}\, Q_{jt} + \\
&\sum_{j=1}^{J}\sum_{t=1}^{T}pc_{jt}\, pq_{jt} + \sum_{k=1}^{K}\sum_{c_f=1}^{C}EF_{kc_f}\, W_{kc_f} + \sum_{i_n=1}^{I}\sum_{c_m=1}^{C}EM_{i_nc_m}\, X_{i_nc_m} + \sum_{c_v=1}^{C}\sum_{t=1}^{T}VC_{c_v}\, U_{vc_vt}
\end{aligned} \tag{48}
$$

4.10. Solution Methodology

To solve this model, an approach that can manage a large number of variables is required, and using classical approaches and methodologies is a time-consuming task. Thus, several mathematical techniques were deployed to obtain the optimal solution of the objective function and thus minimize the total costs by focusing on the optimum location-allocation of the blood facilities. To solve the proposed model quickly, the Lingo solver and the Facility Location Problem solver were used to generate the data and obtain the optimal results. Numerical experiments were conducted, and the results are reported in the following section.

4.11. Numerical Experiments

To validate the proposed model, the case of the city of Ansan, South Korea, was studied. In this scenario, two mobile blood facilities (MBF_1 and MBF_2), two fixed blood facilities (FBF_1 and FBF_2), one blood center (BC), seven hospitals, and five delivery vehicles were considered, as presented in Table 2.

Table 2. Size of the problem.

MBF	FBF	BC	Hospitals	Vehicles
2	2	1	7	5

The model was run on Lingo version 17.0 on an Intel(R) Core(TM) i7-3770 CPU@3.40 GHz 3.40 GHz with a 64-bit operating system and 8 GB of RAM. The determination of candidate locations that can be visited by an MBF during a planning period is a strategic decision that should be made carefully in order to meet the daily demand of blood platelets. Thus, to ensure optimal distribution, location-allocation of the mobile blood facilities was assessed according to the population of each region, the location of hospitals, and historical data. The locations are defined by geographic coordinates (longitude and latitude). The FLP solver with Bing Maps was used to calculate the actual distances between candidate locations in the network and determine the optimal location for each mobile and fixed blood facility. For MBFs, each candidate location can be visited only once. Specifying the optimal locations at which to establish a blood facility is a critical decision and includes the consideration of many conditions, such as the population of each region, the location of hospitals, and the history of donation in each region. Table 3 shows the candidate locations of the blood facilities (MBFs and FBFs) and the fixed locations of the blood center and hospitals in the city of Ansan, South Korea.

Table 3. Candidate locations of blood facilities and the fixed locations of the blood center and hospitals in Ansan, South Korea.

Candidate Locations of the Blood Facilities (MBFs and FBFs).					
ID	Location	Address	Latitude (x)	Longitude (y)	Demand
1	L1	Sangnok-gu, Ansan, South Korea	37.30	126.84	MBF/FBF
2	L2	Sa-dong, Ansan, South Korea	37.29	126.85	MBF/FBF
3	L3	Seonbu-dong, Ansan, South Korea	37.33	126.80	MBF/FBF
4	L4	Choji-dong, Ansan, South Korea	37.32	126.80	MBF/FBF
5	L5	Wongok-dong, Ansan, South Korea	37.32	126.79	MBF/FBF
6	L6	Singil-dong, Ansan, South Korea	37.33	126.76	MBF/FBF
7	L7	Wolpi-dong, Ansan, South Korea	37.33	126.84	MBF/FBF
8	L8	Gojan-dong, Ansan, South Korea	37.32	126.81	MBF/FBF
9	L9	Bono-dong, Ansan, South Korea	37.28	126.86	MBF/FBF
10	L10	Wonsi-dong, Ansan, South Korea	37.30	126.79	FBF
11	L11	Mongnae-dong, Ansan, South Korea	37.30	126.77	FBF
12	L12	Wa-dong, Ansan, South Korea	37.34	126.82	FBF
13	L13	Hosu-dong, Ansan, South Korea	37.32	126.77	FBF
14	L14	Seongpo-dong, Ansan, South Korea	37.32	126.84	FBF
Location of the blood center in Ansan, South Korea.					
ID	Location	Address		Latitude (x)	Longitude (y)
15	L15	Ansan blood center, Ansan, South Korea		37.31	126.87
Locations of the hospitals at Ansan, South Korea					
ID	Location	Address		Latitude (x)	Longitude (y)
1	H1	Danwon Hospital, Choji-dong, Ansan, South Korea		37.32	126.80
2	H2	Korea University Ansan Hospital, 69 516 Gojan-dong, Danwon-gu, Ansan-si, Gyeonggi-do, South Korea		37.32	126.81
3	H3	Duson Hospital, Seonbu-dong, Ansan-si, South Korea		37.33	126.80
4	H4	Sarang Hospital, 69 Yesulgwangjang-ro, Seongpo-dong, Sangnok-gu, Ansan-si,Gyeonggi-do, South Korea		37.32	126.84
5	H5	Hansarang Hospital, 345 Gwangdeok 1 (il)-ro, I-dong, Sangnok-gu, Ansan-si, Gyeonggi-do, South Korea		37.30	126.84
6	H6	Anshan 21st Century Hospital, 50 Bohwa-ro, Gojan 1 (il)-dong, Danwon-gu, Ansan-si, Gyeonggi-do, South Korea		37.33	126.83
7	H7	Ansan Hospital, Korea Labor Welfare Corporation, 95 Il-dong, Sangnok-gu, Ansan-si, Gyeonggi-do, South Korea		37.30	126.86

A map is presented in Figure 3 to visualize the candidate locations of the mobile blood facilities and the fixed blood facilities, as well as the fixed location of the blood center.

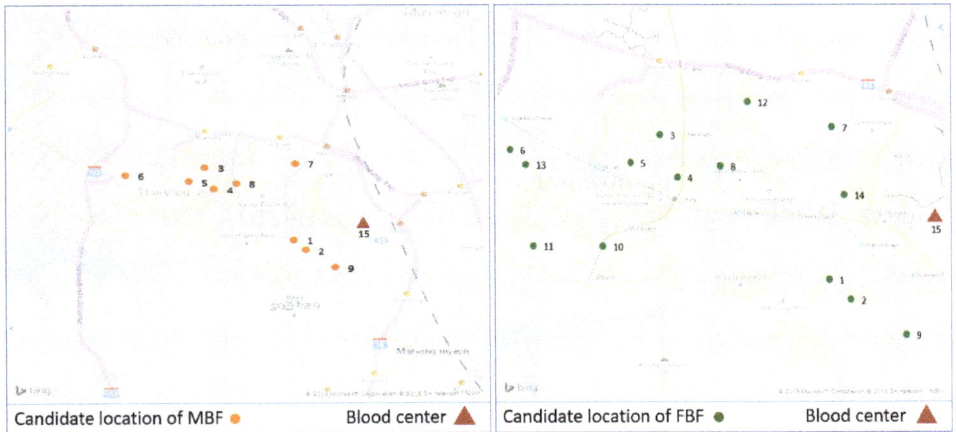

Figure 3. Candidate locations of mobile and fixed blood facilities.

Analyses were carried out to determine the optimal locations of the fixed blood facilities, and the results indicate that locations 1 and 5 are the best sites for establishing fixed blood facilities as shown by Figure 4.

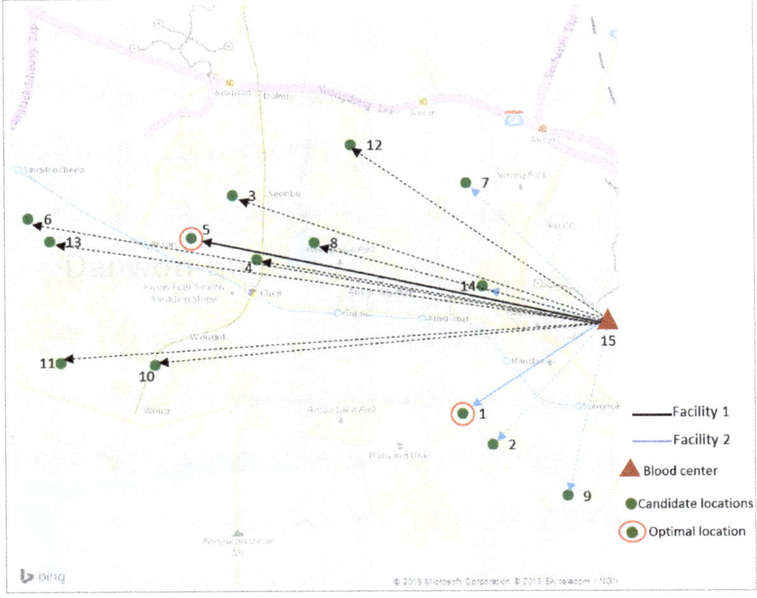

Figure 4. Optimal locations of the fixed blood facilities.

Table 4 presents details of the FBF1 and FBF2 locations and their distances from the blood center. The optimum locations of the fixed blood facilities (FBF1 and FBF2) prove to be location 1 and location 5. Consequently, no mobile blood facilities are allowed to visit these locations because the maximum capacity of each candidate location is one facility, either fixed or mobile. Since the locations of the two fixed blood facilities are determined, mobile blood facilities cannot be assigned to these locations.

Table 4. Candidate and optimal locations of the fixed blood facilities.

Facilities	Candidate Locations	Optimal Location	Distance Between BC and FBF (km)
FBF1	**Location 1**	**Location 1**	5.5
	Location 2		
	Location 7		
	Location 9		
	Location 14		
FBF2	**Location 5**	**Location 5**	9.3
	Location 3		
	Location 4		
	Location 6		
	Location 8		
	Location 10		
	Location 11		
	Location 12		
	Location 13		

The results of analyses reveal that the optimal location for the first mobile blood facility (MBF1) is location 3. In the next planning period, MBF1 can move from location 3 to locations 4, 6, 7, and 8, in that order. An MBF does not have to visit all candidate locations if the required quantity of platelets is met or if the working hours of the employees are over. Once the required quantity of blood platelets is collected, a mobile blood facility can deliver the collected blood platelets directly to the assigned blood center. Table 5 represents the details related to the allocation of the first mobile blood facility, MBF1.

Table 5. The optimal location and navigation path of MBF1.

MBF1	Optimal Location	Address	Candidate Locations	Possible Visited Locations
1	**Location 3**	Seonbu-dong, Ansan, South Korea	9	6

No.	Path	Distance (km)	Demand	Transportation Costs ($)
1	BC to Location 3	9.4	1	5.64
2	Location 3 to Location 4	1.5	1	0.88
3	Location 4 to Location 6	5	1	3
5	Location 6 to Location 7	9.1	1	5.47
6	Location 7 to Location 8	3.2	1	1.91
7	Location 8 to BC	7	-	4.2

The optimal location of the second mobile blood facility (MBF2) is determined to be location 2. MBF2 can move from location 2 to location 9 in the next planning period. Similar to MBF1, if the needed quantity of platelets is collected or if the working hours of the employees are over, MBF2 moves to the designated blood center to deliver the collected platelets. Table 6 represents the details related to the allocation of the second mobile blood facility, MBF2.

Table 6. The optimal location and navigation path of MBF2.

MBF2	Optimal Location	Address	Candidate Locations	Possible Visited Locations
1	**Location 2**	Sa-dong, Ansan, South Korea	9	3

No.	Path	Distance (km)	Demand	Transportation Costs ($)
1	BC to Location 2	5	1	3
2	Location 2 to Location 9	2.4	1	1.44
4	Location 9 to BC	3.5	-	2.1

The optimal locations and navigation paths of the mobile blood facilities (MBF1 and MBF2) are presented in Figure 5.

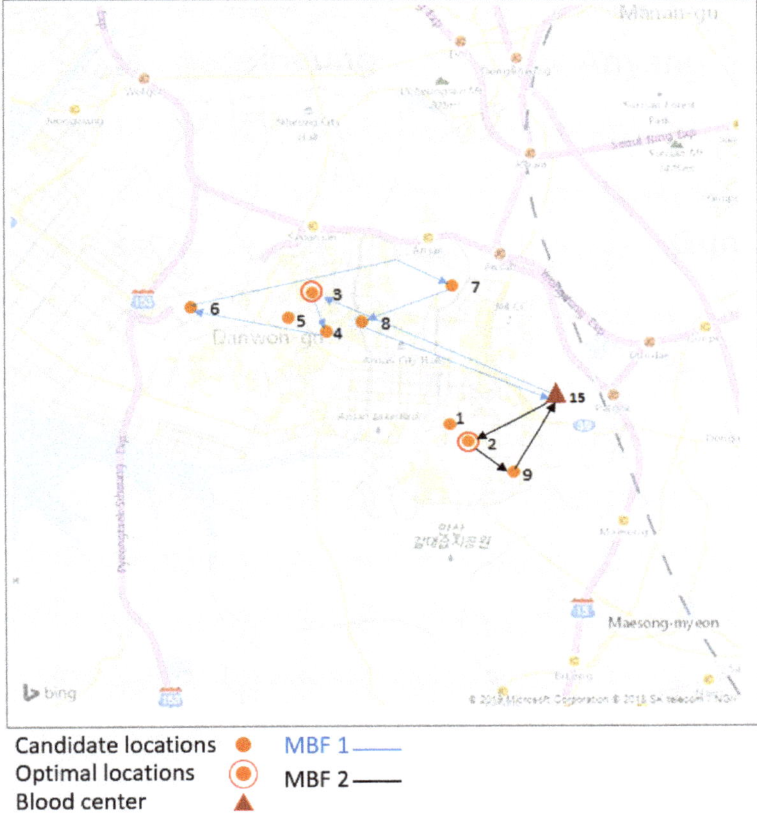

Figure 5. Optimal locations of mobile blood facilities 1 and 2.

In the proposed problem, seven hospitals must be served each day to fulfill their daily demand for blood platelets. The results show that three different routes can be considered to guarantee efficient supply chain management while ensuring optimal energy consumption. In the first route (R1), three hospitals (h1, h2, and h3) must be served by one vehicle. Two hospitals need to be visited in the second route (R2), and two hospitals are also served in the third route (R3). The locations of the hospitals are fixed, and the calculated distances between all nodes of the routes are reported in Table 7.

Table 7. The total distance of each route.

Route	Path	Locations	Distance (km)	Total Distance (km)
R1	BC to h1	L15 to L7	2.4	14.8
	h1 to h2	L7 to L5	4	
	h2 to h3	L5 to L4	2.6	
	h3 to BC	L4 to L15	5.8	
R2	BC to h1	L15 to L6	5.9	13.9
	h1 to h2	L6 to L2	1.6	
	h2 to BC	L2 to L15	6.4	
R3	BC to h1	L15 to L1	7.6	19.3
	h1 to h2	L1 to L3	4.5	
	h2 to BC	L3 to L15	7.2	

The optimum shipping routes represent the best balance between costs and energy consumption. The shipping routes determined to be optimal are presented in Figure 6.

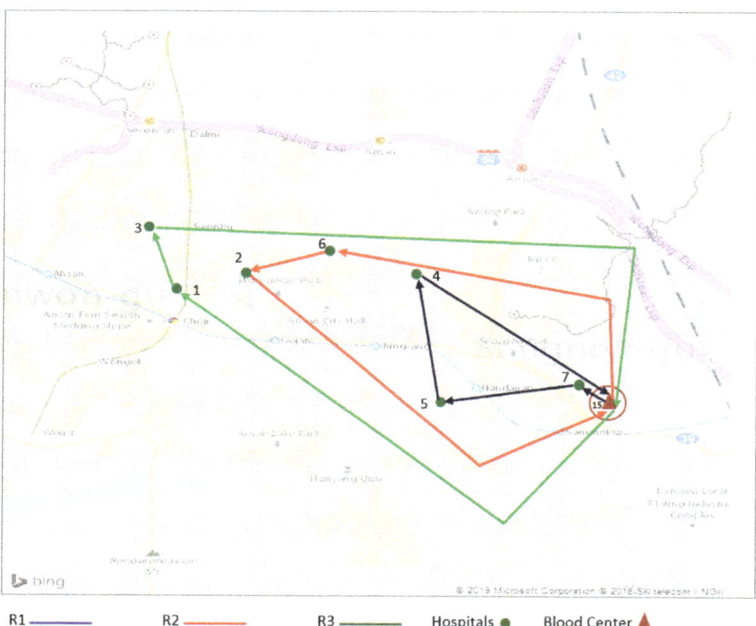

Figure 6. Optimal routes between hospitals and the blood center.

The values of the parameters used in the test problems are presented in Table 8.

Numerical experiments were carried out for different delivery scenarios to validate the proposed model. Ten scenarios were considered in this study to determine the best shipping route and the best scenario with the most efficient results in terms of costs and energy consumption. Table 9 shows the obtained results for the different tested scenarios.

Table 8. Values of the parameters in the test problem (for references, see Khanghahi et al. [17], Heidari et al. [64], Hamdan and Diabat [18], Arvan et al. [7], and Samani et al. [16]).

Parameters	Values	Parameters	Values
c_m	[300, 400]	$APHc_{jt}$	[1, 2]
c_f	[1000, 2000]	EF_{kc_f}	[20,000, 40,000]
c_b	[3000, 8000]	$EM_{i_n c_m}$	[15,000, 30,000]
c_v	[300, 400]	$VC_{c_v t}$	[10,000, 15,000]
Tc	[0.5, 0.7]	Qce	[60, 150]
L	[5, 7]	d_{kj}	[1, 20]
CE_c	4×10^{-5}	d_{i_njt}	[1, 20]
hc_{ajt}	$[14 \times 10^{-3}, 18 \times 10^{-3}]$	d_{jh_it}	[1, 60]
hc_{akt}	$[14 \times 10^{-3}, 18 \times 10^{-3}]$	d_{i_{n-1}i_n}	[0, 10]
α	[0.01, 0.05]	β	[0.1, 0.2]
λ	[0.1, 0.2]	pc_{jt}	[0.5, 0.8]
Cu	[0.015, 0.018]	$APHc_{i_n t}$	[1, 2]
Uc	[0.01, 0.02]	Dm_{jt}	[1, 210]
$APHc_{kt}$	[1, 2]	$Dm_{h_i t}$	[1, 30]

Table 9. Results of different delivery scenarios.

Facilities	Scenario 1A	Scenario 2A	Scenario 3A	Scenario 4A	Scenario 5A
MBF1	BC to L3 to BC	BC to L3 to L4 to BC	BC to L3 to L4 to L6 to BC	BC to L3 to L4 to L6 to L7 to BC	BC to L3 to L4 to L6 to L7 to L8 to BC
		Scenario 1B		Scenario 2B	
MBF2		BC to L2 to BC		BC to L2 to L9 to BC	
FBF1		Scenario 1C: BC to L1 to BC			
FBF2		Scenario 1D: BC to L5 to BC			
Routes	R1: BC to L7' to L5' to L4' to BC		R2: BC to L6' to L2' to BC		R2: BC to L1' to L3' to BC

		TEST	
Scenarios	Runs	Path	Total costs ($/cycle)
Scenario 1	Run 1	Scenario 1A + Scenario 1B + Scenario 1C + Scenario 1D + R1 +R2 +R3	145245.37
Scenario 2	Run 2	Scenario 2A + Scenario 1B + Scenario 1C + Scenario 1D + R1 +R2 +R3	144787.76
Scenario 3	Run 3	Scenario 3A + Scenario 1B + Scenario 1C + Scenario 1D + R1 +R2 +R3	145746.94
Scenario 4	Run 4	Scenario 4A + Scenario 1B + Scenario 1C + Scenario 1D + R1 +R2 +R3	146874.67
Scenario 5	Run 5	Scenario 5A + Scenario 1B + Scenario 1C + Scenario 1D + R1 +R2 +R3	147379.21
Scenario 6	Run 6	Scenario 1A + Scenario 2B + Scenario 1C + Scenario 1D + R1 +R2 +R3	145188.31
Scenario 7 *	**Run 7 ***	**Scenario 2A + Scenario 2B + Scenario 1C + Scenario 1D + R1 +R2 +R3**	**144687.69 ***
Scenario 8	Run 8	Scenario 3A + Scenario 2B + Scenario 1C + Scenario 1D + R1 +R2 +R3	145471.87
Scenario 9	Run 9	Scenario 4A + Scenario 2B + Scenario 1C + Scenario 1D + R1 +R2 +R3	146793.61
Scenario 10	Run 10	Scenario 5A + Scenario 2B + Scenario 1C + Scenario 1D + R1 +R2 +R3	147196.15

* represents the indicator of the optimum values.

Scenario 7 represents the optimal solution to the proposed model, with total costs equal to $144,687.69. The first mobile blood facility (MBF1) has to visit location 3 first, as it was concluded to be its optimal location. Then, it moves to location 4, collects the required quantity of blood platelets to satisfy the daily demand, and returns to the blood center, where the trip started, to deliver the collected quantity of blood platelets. The second mobile blood facility (MBF2) starts at its optimal site of location 2, moves to location 9 in the next planning period, and then returns to the assigned blood center with the collected quantity of blood platelets. The fixed blood facilities (FBF1 and FBF2) reside at their optimal locations, location 1 and location 5, respectively. The blood platelets collected at FBF1 and FBF2 are transported to the blood center. Donors can donate platelets in the blood center as well; thus, the blood center acts not only as a testing, production, storage, and diffusion facility but also as a potential collection site of more blood platelets.

The final quantity of blood platelets produced is determined by the summation of all collected platelets while accounting for the wastage rate and the outdate rate of platelets in time period t. The demand volume is delivered to demand points by three different routes. In the first route, three hospitals are visited; in the second route, two hospitals are visited; and in the third route, two hospitals are served. Thus, the optimum shipping route is given by scenario 7, which represents the best shipping route in terms of efficient energy consumption and minimum costs.

5. Conclusions

This study was conducted to identify an optimal delivery approach by accounting for the effect of optimal energy consumption using objective functions to reduce the total costs of the network. The proposed strategy is demonstrated for the city of Ansan in South Korea. Energy efficiency plays a key role in the strategy to reduce the costs of healthcare supply chain management. The total cost was minimized with the optimal energy consumption, and an optimal transportation scheme for blood platelet supply chain management was developed. A mathematical model was designed with the objective of enhancing the efficiency of healthcare service. The decision-making related to the location-allocation of blood facilities plays an important role in reducing the energy consumed by transportation activities between all nodes in the network. Thus, the total cost was minimized to assure the sustainability of the healthcare supply chain. As a result of the high perishability rate, platelets are a life-saving product in constant demand, and managing the healthcare supply chain is both a major priority and a complicated task. The proposed formulation ensures the optimal location-allocation of mobile and fixed blood facilities. The model determines the optimal path to ensure the efficiency of the supply chain, the reduction of energy consumption, and the efficient flow of blood platelets at required quantities through the network. Further, an objective of the model is the minimization of the total cost of the supply chain (establishment cost of FBFs, setup cost of MBFs and delivery vehicles, production costs, operation costs, transportation costs due to the effect of energy, carbon emission costs, perishability costs, holding costs, and wastage costs). The main output of this model, with respect to the optimal energy, is optimized healthcare supply chain management with a minimized total cost. A set of different scenarios were considered to determine the optimal solution and the best scheme to deliver blood platelets. In addition, data from a real case study was applied to prove the validity of the proposed model. To the best of the authors' knowledge, this is the first study that combines the concept of energy consumption and healthcare supply chain management .

Because of the complexity of the formulated problem, potential future research includes developing efficient solution algorithms that can solve the problem within a reasonable time. The model can be enhanced by adding more objective functions to be minimized or maximized to foster and boost the efficiency of the healthcare supply chain. In future research, the outdating of blood platelet units can be considered as a deterioration linear function or higher-order polynomial, which could allow the model to give priority to younger units of platelets over the old units.

Author Contributions: Conceptualization, J.J. and B.S.; methodology, J.J. and B.S.; software, J.J.; validation, J.J. and B.S.; formal analysis, J.J.; investigation, J.J. and B.S.; resources, J.J. and B.S.; data curation, J.J. and B.S.; writing–original draft preparation, J.J.; writing–review and editing, B.S.; visualization, J.J. and B.S.; supervision, B.S.; project administration, B.S.; funding acquisition, B.S.

Funding: This work was supported by the research fund of Hanyang University (HY-2019-N) (Project number: 201900000000419).

Abbreviations

The following abbreviations were used in this paper:

MBF   Mobile Blood Facility
FBF   Fixed Blood Facility
BC   Blood Cenetr

Appendix A

Sets

I	mobile blood facilities, $i_n \in I$
K	fixed blood facilities, $k \in K$
J	blood center, $j \in J$
H	hospitals, $h_i \in H$
V	vehicles indexed by $v_i \in V$
C	capacities $c_i \in C$
T	planning periods, $t \in T$
R	routes indexed by $r_i \in R$ with $i = 1, \cdots, n$
A	age of platelets, $a \in A$

Parameters

c_i capacity with

$$
i = \begin{cases}
v: & \text{vehicle} \\
b: & \text{blood center} \\
f: & \text{fixed blood facility} \\
vm: & \text{mobile blood facility}
\end{cases}
$$

Tc	fixed transportation costs under efficient-energy consumption per km ($/km)
L	lifetime of platelets (days)
CE_c	carbon emission costs ($/gr of CO_2)
hc_{ajt}	holding costs of the blood with age a in the blood center during the planning period t ($/unit/unit time)
hc_{akt}	holding costs at fixed blood facilities during the planning period t ($/unit/unit time)
π	penalty costs for shortage in blood age a at the blood center during the planning period t ($/unit/unit time)
Cu	unit cost of expired blood at a blood center ($/unit/unit time)
Uc	cost of one expired blood unit at a fixed blood facility ($/unit/unit time)
pc_{jt}	unit production costs of platelets at blood center ($/unit)
$APHc_{i_n t}$	operational cost of platelets at mobile blood facility ($/unit)
$APHc_{kt}$	operating cost of platelets at fixed blood facility ($/unit)
$APHc_{jt}$	cost of operating the apheresis machine at blood center ($/unit)
EF_{kc_f}	establishment costs of a fixed blood facility ($/setup)
$EM_{i_n c_m}$	establishment costs of a mobile blood facility ($/setup)
$VC_{c_v t}$	setup costs of delivery vehicle with capacity c_v ($/setup)
Qce	quantity of carbon emitted by the vehicle per km (gr of CO_2 / km)
d_{kj}	distance between a fixed blood facility and blood center (km)
$d_{i_n jt}$	total distance a mobile blood facility travel during the planning period t to collect the platelets (km) and deliver them to the blood center (km)
$d_{jh_i t}$	total distance a vehicle travel during the planning period t in order to diffuse the platelets to all the demand points (km)
$d_{i_{n-1} i_n}$	distance between location i_{n-1} and location i_n (km)
α	wasted blood platelets rate at blood center during production process
β	outdated blood platelets rate at blood center
λ	expired blood platelets rate at a fixed blood facility
$Dm_{h_i t}$	quantity of demand of hospitals (unit/unit time)
Dm_{jt}	demand of blood center (unit/unit time)
M	very large number

Decision variables

Q_{kjt}	transported quantity of blood from a fixed blood facility to a blood center in period t (unit/unit time)
Q_{kt}	collected quantity of blood in a fixed blood facility in period t (unit/unit time)
Q_{i_njt}	shipped quantity of blood from a mobile blood facility to a blood center (unit/unit time)
Q_{jt}	collected quantity of blood in a blood center in period t (unit/unit time)
Q_{jh_it}	quantity of blood to be transported from the blood center to the demand points (unit/unit time)
Qs_{ajt}	shortage quantity of blood age a at the blood center during the planning period t (unit/unit time)
Qs_{ajt}^{max}	maximum quantity of shortage allowed (unit/unit time)
Qw_t	wasted quantity of blood (unit/unit time)
Qu_t	expired quantity of blood at blood center (unit/unit time)
Qp_t	quantity of expired blood units at fixed blood facility (unit/unit time)
pq_{jt}	produced quantity of blood at the blood center during the planning period t (unit/unit time)
IL_{akt}	inventory level at a fixed blood facility during the planning period t (unit/unit time)
IL_{ajt}	inventory level at a blood center during the period t (unit/unit time)

Binary variables

$$U_{vc_vt} = \begin{cases} 1 & \text{if a vehicle } v \text{ with capacity } c_v \text{ is used during the time period } t \\ 0; & \text{otherwise} \end{cases}$$

$$W_{kc_f} = \begin{cases} 1 & \text{if a fixed blood center with capacity } c_f \text{ is established at location } k \\ 0; & \text{otherwise} \end{cases}$$

$$X_{i_nc_m} = \begin{cases} 1 & \text{if a mobile blood facility with capacity } c_m \text{ is established at location } i_n \\ 0; & \text{otherwise} \end{cases}$$

$$Y_{jc_b} = \begin{cases} 1 & \text{if a blood center with capacity } c_b \text{ is open} \\ 0; & \text{otherwise} \end{cases}$$

$$M_{c_mi_njt} = \begin{cases} 1 & \text{if a mobile blood facility is assigned to a blood center during the planning period } t \\ 0; & \text{otherwise} \end{cases}$$

$$F_{kjt} = \begin{cases} 1 & \text{if a fixed blood facility is assigned to a blood center during the time period } t \\ 0; & \text{otherwise} \end{cases}$$

$$B_{jh_it} = \begin{cases} 1 & \text{if a blood center is assigned to a hospital during the time period } t \\ 0; & \text{otherwise} \end{cases}$$

$$H_{h_ijt} = \begin{cases} 1 & \text{if a hospital is assigned to a blood center during the planning period } t \\ 0; & \text{otherwise} \end{cases}$$

$$G_{i_{n-1}i_nt} = \begin{cases} 1 & \text{if a mobile blood facility is assigned to location } i_{n-1} \text{ in the planning period } t-1 \text{ and} \\ & \text{moves to location } i_n \text{ in the next planning period } t \\ 0; & \text{otherwise} \end{cases}$$

References

1. Holinstat, M. Normal platelet function. *Cancer Metastasis Rev.* **2017**, *36*, 195–198. [CrossRef] [PubMed]
2. Nahmias, S. Perishable inventory theory: A review. *Oper. Res.* **1982**, *30*, 680–708. [CrossRef] [PubMed]
3. Sha, Y.; Huang, J. The multi-period location-allocation problem of engineering emergency blood supply systems. *Syst. Eng. Procedia* **2012**, *5*, 21–28. [CrossRef]
4. Jabbarzadeh, A.; Fahimnia, B.; Seuring, S. Dynamic supply chain network design for the supply of blood in disasters: A robust model with real world application. *Trans. Res. Part E Logist. Transp. Rev.* **2014**, *70*, 225–244. [CrossRef]
5. Katsaliaki, K.; Mustafee, N.; Kumar, S. A game-based approach towards facilitating decision making for perishable products: An example of blood supply chain. *Exp. Syst. Appl.* **2014**, *41*, 4043–4059. [CrossRef]
6. Duan, Q.; Liao, T.W. Optimization of blood supply chain with shortened shelf lives and ABO compatibility. *Int. J. Prod. Econ.* **2014**, *153*, 113–129. [CrossRef]

7.   Arvan, M.; Tavakkoli-Moghaddam, R.; Abdollahi, M. Designing a bi-objective and multi-product supply chain network for the supply of blood. *Uncertain Supply Chain Manag.* **2015**, *3*, 57–68. [CrossRef]

8.   Giannakis, M.; Papadopoulos, T. Supply chain sustainability: A risk management approach. *Int. J. Prod. Econ.* **2016**, *171*, 455–470. [CrossRef]

9.   Osorio, A.F.; Brailsford, S.C.; Smith, H.K.; Forero-Matiz, S.P.; Camacho-Rodríguez, B.A. Simulation-optimization model for production planning in the blood supply chain. *Healthc. Manag. Sci.* **2017**, *20*, 548–564. [CrossRef]

10.   Zahiri, B.; Pishvaee, M.S. Blood supply chain network design considering blood group compatibility under uncertainty. *Int. J. Prod. Res.* **2017**, *55*, 2013–2033. [CrossRef]

11.   Ramezanian, R.; Behboodi, Z. Blood supply chain network design under uncertainties in supply and demand considering social aspects. *Trans. Res. Part E Logist. Transp. Rev.* **2017**, *104*, 69–82. [CrossRef]

12.   Paydar, M.M.; Babaveisi, V.; Safaei, A.S. An engine oil closed-loop supply chain design considering collection risk. *Comput. Chem. Eng.* **2017**, *104*, 38–55. [CrossRef]

13.   Najafi, M.; Ahmadi, A.; Zolfagharinia, H. Blood inventory management in hospitals: Considering supply and demand uncertainty and blood transshipment possibility. *Oper. Res. Health Care* **2017**, *15*, 43–56. [CrossRef]

14.   Ensafian, H.; Yaghoubi, S. Robust optimization model for integrated procurement, production and distribution in platelet supply chain. *Trans. Res. Part E Logist. Transp. Rev.* **2017**, *103*, 32–55. [CrossRef]

15.   Hosseinifard, Z.; Abbasi, B. The inventory centralization impacts on sustainability of the blood supply chain. *Comput. Oper. Res.* **2018**, *89*, 206–212. [CrossRef]

16.   Samani, M.R.G.; Torabi, S.A.; Hosseini-Motlagh, S.M. Integrated blood supply chain planning for disaster relief. *Int. J. Disaster Risk Reduct.* **2018**, *27*, 168–188. [CrossRef]

17.   Eskandari-Khanghahi, M.; Tavakkoli-Moghaddam, R.; Taleizadeh, A.A.; Amin, S.H. Designing and optimizing a sustainable supply chain network for a blood platelet bank under uncertainty. *Eng. Appl. Artif. Intell.* **2018**, *71*, 236–250. [CrossRef]

18.   Hamdan, B.; Diabat, A. A two-stage multi-echelon stochastic blood supply chain problem. *Comput. Oper. Res.* **2019**, *101*, 130–143. [CrossRef]

19.   Beamon, B.M. Sustainability and the future of supply chain management. *Oper. Supply Chain Manag.* **2008**, *1*, 4–18. [CrossRef]

20.   Rentizelas, A.A.; Tolis, A.J.; Tatsiopoulos, I.P. Logistics issues of biomass: The storage problem and the multi-biomass supply chain. *Renew. Sustain. Energy Rev.* **2009**, *13*, 887–894. [CrossRef]

21.   Halldórsson, Á.; Kovács, G. The sustainable agenda and energy efficiency: Logistics solutions and supply chains in times of climate change. *Int. J. Phys. Distrib. Logist. Manag.* **2010**, *40*, 5–13. [CrossRef]

22.   Gold, S.; Seuring, S. Supply chain and logistics issues of bio-energy production. *J. Clean. Prod.* **2011**, *19*, 32–42. [CrossRef]

23.   Pan, S.Y.; Du, M.A.; Huang, I.T.; Liu, I.H.; Chang, E.; Chiang, P.C. Strategies on implementation of waste-to-energy (WTE) supply chain for circular economy system: A review. *J. Clean. Prod.* **2015**, *108*, 409–421. [CrossRef]

24.   Luthra, S.; Mangla, S.K.; Xu, L.; Diabat, A. Using AHP to evaluate barriers in adopting sustainable consumption and production initiatives in a supply chain. *Int. J. Prod. Econ.* **2016**, *181*, 342–349. [CrossRef]

25.   Hong, J.; Shen, Q.; Xue, F. A multi-regional structural path analysis of the energy supply chain in China's construction industry. *Energy Policy* **2016**, *92*, 56–68. [CrossRef]

26.   Marzband, M.; Ghazimirsaeid, S.S.; Uppal, H.; Fernando, T. A real-time evaluation of energy management systems for smart hybrid home Microgrids. *Electr. Power Syst. Res.* **2017**, *143*, 624–633. [CrossRef]

27.   Aziziankohan, A.; Jolai, F.; Khalilzadeh, M.; Soltani, R.; Tavakkoli-Moghaddam, R. Green supply chain management using the queuing theory to handle congestion and reduce energy consumption and emissions from supply chain transportation fleet. *J. Ind. Eng. Manag.* **2017**, *10*, 213–236. [CrossRef]

28.   Centobelli, P.; Cerchione, R.; Esposito, E. Environmental sustainability and energy-efficient supply chain management: A review of research trends and proposed guidelines. *Energies* **2018**, *11*, 275. [CrossRef]

29.   Singh, R.; Centobelli, P.; Cerchione, R. Evaluating Partnerships in Sustainability-Oriented Food Supply Chain: A Five-Stage Performance Measurement Model. *Energies* **2018**, *11*, 3473.

30.   Sarkar, M.; Sarkar, B.; Iqbal, M. Effect of Energy and Failure Rate in a Multi-Item Smart Production System. *Energies* **2018**, *11*, 2958. [CrossRef]

31. Sarkar, B.; Omair, M.; Choi, S.B. A multi-objective optimization of energy, economic, and carbon emission in a production model under sustainable supply chain management. *Appl. Sci.* **2018**, *8*, 1744. [CrossRef]

32. Marchi, B.; Zanoni, S.; Ferretti, I.; Zavanella, L. Stimulating investments in energy efficiency through supply chain integration. *Energies* **2018**, *11*, 858. [CrossRef]

33. Fontes, C.H.d.O.; Freires, F.G.M. Sustainable and renewable energy supply chain: A system dynamics overview. *Renew. Sustain. Energy Rev.* **2018**, *82*, 247–259.

34. Zhu, Q.; Sarkis, J.; Lai, K.h. Green supply chain management: Pressures, practices and performance within the Chinese automobile industry. *J. Clean. Prod.* **2007**, *15*, 1041–1052. [CrossRef]

35. Srivastava, S.K. Green supply-chain management: A state-of-the-art literature review. *Int. J. Manag. Rev.* **2007**, *9*, 53–80. [CrossRef]

36. Yeh, W.C.; Chuang, M.C. Using multi-objective genetic algorithm for partner selection in green supply chain problems. *Exp. Syst. Appl.* **2011**, *38*, 4244–4253. [CrossRef]

37. Hua, G.; Cheng, T.; Wang, S. Managing carbon footprints in inventory management. *Int. J. Prod. Econ.* **2011**, *132*, 178–185. [CrossRef]

38. Kumar, S.; Chattopadhyaya, S.; Sharma, V. Green supply chain management: A case study from Indian electrical and electronics industry. *Int. J. Soft Comput. Eng.* **2012**, *1*, 275–281.

39. Pan, S.; Ballot, E.; Fontane, F. The reduction of greenhouse gas emissions from freight transport by pooling supply chains. *Int. J. Prod. Econ.* **2013**, *143*, 86–94. [CrossRef]

40. Bazan, E.; Jaber, M.Y.; Zanoni, S. Supply chain models with greenhouse gases emissions, energy usage and different coordination decisions. *Appl. Math. Model.* **2015**, *39*, 5131–5151. [CrossRef]

41. Ahmed, W.; Sarkar, B. Impact of carbon emissions in a sustainable supply chain management for a second generation biofuel. *J. Clean. Prod.* **2018**, *186*, 807–820. [CrossRef]

42. Sarkar, B. A production-inventory model with probabilistic deterioration in two-echelon supply chain management. *Appl. Math. Model.* **2013**, *37*, 3138–3151. [CrossRef]

43. Sarkar, B.; Ganguly, B.; Sarkar, M.; Pareek, S. Effect of variable transportation and carbon emission in a three-echelon supply chain model. *Trans. Res. Part E Logist. Transp. Rev.* **2016**, *91*, 112–128. [CrossRef]

44. Habib, M.S.; Sarkar, B. An integrated location-allocation model for temporary disaster debris management under an uncertain environment. *Sustainability* **2017**, *9*, 716. [CrossRef]

45. Sarkar, B.; Shaw, B.K.; Kim, T.; Sarkar, M.; Shin, D. An integrated inventory model with variable transportation cost, two-stage inspection, and defective items. *J. Ind. Manag. Opt.* **2017**, *13*, 1975–1990. [CrossRef]

46. Feng, X.; Moon, I.; Ryu, K. Warehouse capacity sharing via transshipment for an integrated two-echelon supply chain. *Trans. Res. Part E Logist. Transp. Rev.* **2017**, *104*, 17–35. [CrossRef]

47. Sarkar, B.; Ahmed, W.; Kim, N. Joint effects of variable carbon emission cost and multi-delay-in-payments under single-setup-multiple-delivery policy in a global sustainable supply chain. *J. Clean. Prod.* **2018**, *185*, 421–445. [CrossRef]

48. Shi, X.; Zhang, X.; Dong, C.; Wen, S. Economic performance and emission reduction of supply chains in different power structures: Perspective of sustainable investment. *Energies* **2018**, *11*, 983. [CrossRef]

49. Iqbal, M.W.; Sarkar, B. Recycling of lifetime dependent deteriorated products through different supply chains. *RAIRO Oper. Res.* **2019**, *53*, 129–156. [CrossRef]

50. Mishra, U.; Wu, J.Z.; Chiu, A.S.F. Effects of Carbon-Emission and Setup Cost Reduction in a Sustainable Electrical Energy Supply Chain Inventory System. *Energies* **2019**, *12*, 1226. [CrossRef]

51. Sarkar, B.; Mahapatra, A.S. Periodic review fuzzy inventory model with variable lead time and fuzzy demand. *Int. Trans. Oper. Res.* **2017**, *24*, 1197–1227. [CrossRef]

52. Sarkar, B.; Ullah, M.; Kim, N. Environmental and economic assessment of closed-loop supply chain with remanufacturing and returnable transport items. *Comput. Ind. Eng.* **2017**, *111*, 148–163. [CrossRef]

53. Malik, A.I.; Sarkar, B. Coordinating Supply-Chain Management under Stochastic Fuzzy Environment and Lead-Time Reduction. *Mathematics* **2019**, *7*, 480. [CrossRef]

54. Shin, D.; Guchhait, R.; Sarkar, B.; Mittal, M. Controllable lead time, service level constraint, and transportation discounts in a continuous review inventory model. *RAIRO Oper. Res.* **2016**, *50*, 921–934. [CrossRef]

55. Sarkar, B.; Tayyab, M.; Kim, N.; Habib, M.S. Optimal production delivery policies for supplier and manufacturer in a constrained closed-loop supply chain for returnable transport packaging through metaheuristic approach. *Comput. Ind. Eng.* **2019**. [CrossRef]

56. Taleizadeh, A.A.; Babaei, M.S.; Sana, S.S.; Sarkar, B. Pricing Decision within an Inventory Model for Complementary and Substitutable Products. *Mathematics* **2019**, *7*, 568. [CrossRef]

57. Noh, J.; Kim, J.S.; Sarkar, B. Two-echelon supply chain coordination with advertising-driven demand under Stackelberg game policy. *Eur. J. Ind. Eng.* **2019**, *13*, 213–244. [CrossRef]

58. Wook Kang, C.; Ullah, M.; Sarkar, M.; Omair, M.; Sarkar, B. A Single-Stage Manufacturing Model with Imperfect Items, Inspections, Rework, and Planned Backorders. *Mathematics* **2019**, *7*, 446. [CrossRef]

59. Kim, S.J.; Sarkar, B. Supply chain model with stochastic lead time, trade-credit financing, and transportation discounts. *Math. Prob. Eng.* **2017**, *2017*, 6465912. [CrossRef]

60. Sarkar, B.; Tayyab, M.; Choi, S.B. Product Channeling in an O2O Supply Chain Management as Power Transmission in Electric Power Distribution Systems. *Mathematics* **2019**, *7*, 4. [CrossRef]

61. Sarkar, B.; Ullah, M.; Choi, S.B. Joint Inventory and Pricing Policy for an Online to Offline Closed-Loop Supply Chain Model with Random Defective Rate and Returnable Transport Items. *Mathematics* **2019**, *7*, 497. [CrossRef]

62. Bhuniya, S.; Sarkar, B.; Pareek, S. Multi-Product Production System with the Reduced Failure Rate and the Optimum Energy Consumption under Variable Demand. *Mathematics* **2019**, *7*, 465. [CrossRef]

63. Sarkar, B.; Guchhait, R.; Sarkar, M.; Cárdenas-Barrón, L.E. How does an industry manage the optimum cash flow within a smart production system with the carbon footprint and carbon emission under logistics framework? *Int. J. Prod. Econ.* **2019**, *213*, 243–257. [CrossRef]

64. Heidari-Fathian, H.; Pasandideh, S.H.R. Green-blood supply chain network design: Robust optimization, bounded objective function & Lagrangian relaxation. *Comput. Ind. Eng.* **2018**, *122*, 95–105.

Article

An Application of Time-Dependent Holding Costs and System Reliability in a Multi-Item Sustainable Economic Energy Efficient Reliable Manufacturing System

Mitali Sarkar [1], Sungjun Kim [2], Jihed Jemai [3], Baishakhi Ganguly [4] and Biswajit Sarkar [2,*]

[1] Department of Industrial Engineering, Yonsei University, 50 Yonsei-ro, Sinchon-dong, Seodaemun-gu, Seoul 03722, Korea

[2] Department of Industrial & Management Engineering, Hanyang University, Ansan, Gyeonggi-do 155 88, Korea

[3] Department of Industrial Engineering, Hanyang University, Seoul 04763, Korea

[4] Department of Mathematics & Statistics, Banasthali Vidyapith, Rajasthan 304 022, India

* Correspondence: bsbiswajitsarkar@gmail.com; Tel.: +82-10-7498-1981

Received: 2 June 2019; Accepted: 20 July 2019; Published: 25 July 2019

Abstract: Sustainable efficient energy is the key factor of any sustainable manufacturing system. This study addresses a multi-item sustainable economic energy efficient reliable manufacturing quantity (MSEEERMQ) model. The manufacturing system produces defective products during long-runs, where those products may be reworked under the optimum effect of energy and carbon footprint with some costs. As all products are not sold immediately, the holding cost increases based on time. The management decides the system design variable to reduce energy consumption cost and increase system reliability under some time-dependent holding costs, and the optimum energy such that the maximum profit of the production model is obtained with a system reliability as a decision variable. The inflation and time-value of money are considered to calculate the cost of the production model under efficient energy. Using control theory, an Euler–Lagrange method is employed to obtain the sustainable critical path, which gives the optimal solution of the model. There are two lemmas to prove the global optimal solution of the model through the control theory. There is an illustrative example to test the model. Under different conditions there are other two examples with graphical representation and sensitivity analysis. Numerical studies reveal that maximum profit is obtained at the optimal value of the decision variable.

Keywords: energy; sustainable manufacturing system; multi-item production; variable holding cost; inflation; control theory

1. Introduction

In reality, optimization through control theory is a critical challenge for any researcher for any production model under optimum energy. An economic energy efficient reliable manufacturing quantity (EEERMQ) model is one of the challenging research areas (Cárdenas-Barrón et al. [1]). To make a system reliable, the management system of any production industry sector under efficient energy has to follow several steps towards the reliability of machinery systems and products. The sustainable energy efficient economic reliable manufacturing quantity model (SEEERMQ) is the extended version of the EEERMQ with the concept that the type of production quantity is always sustainable, i.e., for any amount of order quantity from the buyer or retailer, the EEERMQ is almost similar in every production system. Thus, to reach the optimum EEERMQ, the production system must be stable. It means that the management needs 100% reliable products, which is achievable when the machinery system under the optimum energy consumption works within their capacity. All labors work properly, support of redundancy machines are available, no defective items are produced or if produced, all defective items are discarded from the system. To make reliable products, the industry has to choose one option in between two possible options such as (i) invest more funds to make the system energy efficient reliable or (ii) without investment, reduce the system failure rate under optimum energy consumption. Sana [2] is the pioneer researcher in this field of study. He introduced the above second condition, that without investment the failure rate can be reduced. He used the concept of development cost of Mettas [3] but both of them did not think about efficient energy. Cao and Schiniederjans [4] examined the cost of inventory and production cost in just-in-time production systems but they considered constant production cost without any energy and carbon footprint efficiency whereas the time-dependent variable production cost is more relevant for any sustainable manufacturing system. Wang [5] developed a free-repair warranty policy for defective products which come with an economic manufacturing system. Giri and Dohi [6] generated the basic economic manufacturing model with a random breakdown and repair, where they considered that time for repair corrective maintenance, and preventive maintenance was a random variable. They considered a variable production rate, but none of them considered any concept of energy consumption or any energy reduction policy. Sarkar et al. [7] extended the model of Giri and Dohi [6] with variable safety stock but with constant production rate and perfect production system without any energy and carbon footprint effect. Sarkar et al. [8] extended their own model with imperfect production with constant production rate, but they did not consider any concept of sustainable energy issue. Sarkar et al. [9] extended Sana's [2] model with a stock-dependent demand pattern. They proved that a stock-dependent demand pattern is the best when imperfect production is considered with constant production rate but they did not think about the sustainability issue. Chiu et al. [10] obtained the best replenishment policy of the economic manufacturing quantity model where defective products and multiple shipments are considered. The production rate was considered in this case. They indicated that just-in-time manufacturing systems can converge over basic economic manufacturing models without any sustainable choice. Sarkar et al. [11] extended the basic economic manufacturing quantity model with a time-dependent production rate, where inflation and time-value of money is considered with constant holding cost. They found that holding cost is very sensitive to the EMQ model. That is the reason for this proposed model to consider time-dependent holding cost. By introducing an advertising cost, a selling-price-dependent demand and a variable holding cost, Sarkar [12] developed an economic manufacturing quantity (EMQ) model with failure rate as a decision variable, but did not think about efficient energy consumption. Sarkar et al. [13] extended the basic EMQ model with time-dependent holding cost, variable demand and variable material cost, which depends on failure rate i.e., less material cost indicates low standard material and high failure. Sarkar and Saren [14] extended Wang's [5] model with inspection errors and warranty costs, but they considered a constant production rate. Omair et al. [15] explained a sustainable production system which can give profits. Jaber et al. [16] first proved that the basic production model is more effective than a just-in-time manufacturing system where entropy cost, specifically energy, effects and worker's stress are added. They proved that just-in-time manufacturing systems can produce more waste than the basic manufacturing model if the shipment is more frequent. They considered this for both EMQ and JIT models, which where effective to reduce holding inventory and others. Ahmed and Sarkar [17] proved that a sustainable production system can be obtained if the production system can be made under triple bottom line approach (like economic, environmental, and social) but they did not consider any effect of energy. Tiwari et al. [18] discussed the sustainable inventory in a basic inventory model, but no one till now has considered the production of multi-products with time-dependent production rate under sustainable manufacturing systems. This model fulfils the gap of multi-product in a SERMQ model where the holding costs are time-dependent and the model is solved using control theory. Recently, Sarkar and Sarkar [19] developed a smart production system with the effect of energy, but the main issue remains the same; using a traditional production system, how can an industry can make a sustainable energy efficient production system? There are several studies (Kluczek [20] and Harris et al. [21], Dehning et al. [22]) considering how smart production solved the issues of efficient energy and all other issues like waste management (Khalil et al. [23], Nižetić et al. [24], Biel and Glock [25]), but not all industries can adopt the smart production system. Therefore, how those industries can solve their issues with efficient energy effects under the traditional manufacturing system is addrssed in this study. Table 1 shows the comparison among different studies in this field of study.

Table 1. Comparison between the contributions of different authors.

Author (s)	Type of Model	Product Type	Energy Efficient (EE) and Reliable (R)	Product Category	Unit Production Cost	Energy
Mettas (2000)	EPQ	P&I	R	Single	Variable	NA
Cao and Schiniederjans (2004)	EPQ	P	NA	Single	Constant	NA
Wang (2004)	EPQ	P&I	NA	Single	Constant	NA
Giri and Dohi (2005)	EMQ	P	R	Single	Constant	NA
Sana (2010)	EMQ	P&I	R	Single	Variable	NA
Sarkar et al. (2010)	EMQ	P	R	Single	Variable	NA
Chiu et al. (2011)	EMQ	P&I	NA	Single	Constant	NA
Sarkar (2012)	EMQ	P&I	R	Single	Variable	NA
Cárdenas-Barrón et al. (2013)						
Sarkar and Saren (2016)	EPQ	P&I	NA	Single	Constant	NA
Jaber et al. (2017)	SEPQ	P	NA	Single	Constant	Energy
Omair et al. (2017)	EPQ	P	NA	Single	Constant	NA
Ahmed and Sarkar (2018)	SEPQ	P	NA	Single	Constant	NA
Tiwari et al. (2018)	EMQ	P	NA	Single	Constant	NA
This Model	MSEEEMQ	P&I	EE&R	Multiple	Variable	Energy

P, I, and NA indicate perfect products, imperfect products, and not applicable, respectively.

From Table 1, it can be concluded that there is a big research gap in multi-item sustainable economic energy efficient reliable manufacturing quantity models in the direction of the effect of energy. There are several studies, but no one has considered the effect of energy. Many existing models considered a constant unit production rate, but in reality it generally depends on the rate of production. In this direction, almost all models developed for single products; this model includes some issues that can contribute to obtain sustainable processes. This proposed research fulfills the research gap.

2. Problem Definition, Notation, and Assumptions

This section consists of problem definition, notation, and assumptions of the model.

2.1. Problem Definition

The aim of the model is to obtain the optimal design variable for a multi-item economic energy efficient reliable manufacturing quantity model with maximum profit. The production system is being made sustainable with the help of system reliability and efficient use of energy. A basic manufacturing system is considered for this purpose. The effect of energy is considered within each sector of the production system. As for long-time holding, the holding cost is never constant. Thus, a time-dependent holding cost under the effect of inflation and time-value of money is considered to calculate the profit. The defective products are produced with a random rate. Hence, excess energy consumptions are there for reworking those defective products into perfect products. To make a sustainable energy efficient production system, the carbon footprint is taken into consideration. The demand pattern should be time-dependent and selling-price-dependent which can give more profit than the unrealistic constant demand. The profit is maximized for the sustainable manufacturing system with the help of the Euler–Lagrange theory under the use of optimal energy.

2.2. Notation

The notation of this model is given in Table 2.

Table 2. Notation for index, variables and parameters.

Index	
i	number of products $i = 1, 2, \ldots, n$
Decision variables	
η	failure rate of the manufacturing system of product i
Parameters	
A	development cost without the energy and carbon footprint cost (\$/unit)
A'	energy and carbon footprint cost for developing product (\$/unit)
B	cost related with system technology and reliability
C_i	inspection cost without energy consumption cost of product i, (\$/unit)
C'_i	energy and carbon footprint cost for inspection of product i, (\$/unit)
C_{d_i}	reworking cost of product i without the energy and carbon footprint cost (\$/unit)
C'_{d_i}	energy and carbon footprint cost for reworking of product i (\$/unit)
C_{p_i}	manufacturing cost without energy and carbon footprint cost of product i (\$/unit)
C'_{p_i}	energy and carbon footprint cost for manufacturing product i (\$/unit)
C_{0_i}	material cost of product i, (\$/unit)
C'_{0_i}	material cost for energy of product i, (\$/unit)
k	scaling parameter of design variable
p_i	selling-price product i (\$/unit)
p_{max}	maximum value of selling price (\$/unit)
p_{min}	minimum value of selling price (\$/unit)
Q_i	lot size of product i, (unit)
T	length of production cycle (year)
τ	inflation rate and time-value of money
α	tool/die cost without energy and carbon footprint cost (\$/unit)
η_{max}	maximum value of failure rate
η_{min}	minimum value of failure rate
Expressions	
C_{h_i}	holding cost with energy effect of product i (\$/unit/unit time), $C_{h_i} = (C_{h_{1_i}} + C'_{h_{1_i}}) + (C_{h_{2_i}} + C'_{h_{2_i}})t$
D_i	demand for product i, $D_i = ae^{bt} + \frac{p_{max} - p_i}{p_i - p_{min}}$
$S_i(t)$	time-dependent production rate under the optimum energy effect for product i
Ω_1	total profit under the effect of energy and carbon footprint per cycle (\$/cycle)
Ω_2	total profit under the effect of energy as well as carbon footprint, inflation, and time-value of money (\$/cycle)
Ω_3	final total profit per cycle of the MSEEEMQ (\$/cycle)

Those above mentioned notation is utilized for the proposed model.

2.3. Assumptions

The following assumptions are considered to develop this model.

1. This model considers a sustainable economic energy efficient reliable manufacturing quantity model (SEEERMQ) for a multi-item in an imperfect production system.

2. During a long-run production process, the system moves to an *out-of-control* state from an *in-control* state and it starts to produce some defective items. The production of defective items is a very small percentage of production rate as the industry manager always maintains a reliable system. For maintaining the brand image of a company, a whole-lot-inspection policy is utilized to sperate

the defective items. The defective items are reworked under optimum energy consumption with a fixed cost to make them as new (Sarkar [26], Sarkar et al. [27], San-Joŝe et al. [28]).

3. For the reliable energy efficient production system, the management of the industry has two choices; invest more funds to increase the system reliability under optimum energy or maintain the whole system in such a way that the failure rate with the optimum energy can be reduced. Therefore, the system design variable for η is defined as follows:

$$\eta = \frac{\text{Numbers of failures}}{\text{Total number of working hours}},\tag{1}$$

i.e., less number of failure indicates a more reliable system. The industry managers use this opportunity to make the system reliable by reducing the failure rate of the production system to make a sustainable EMQ (Sarkar [29], Shin et al. [30], Cárdenas-Barrón et al. [31]).

4. As this is a multi-item energy efficient production system, the holding cost is different for different items depending on time linearly. The demand of a multi-item is assumed as time and selling-price dependent (Manente et al. [32], Govindan et al. [33], Sarkar et al. [34]).

5. The development cost under optimum energy is considered as a function of a system design variable and the corresponding unit production cost of the system under the optimum energy consumption is dependent on the development cost, material cost, and tool/die cost. All costs are related to optimum energy consumption (Sarkar et al. [35], Govindan et al. [36]).

6. The inflation and time-value of money are considered to obtain profits. The time horizon is considered as finite and the initial inventory, as well as the final inventory, are zero at the initial and final boundary points.

3. Optimization Problem Development

This section contains a mathematical model, solution of the optimization problem and verification of the optimality conditions.

3.1. Mathematical Model

To make a multi-item sustainable economic energy efficient reliable manufacturing quantity (MSEEERMQ) model, it is essential to make the manufacturing system sustainable and reliable under efficient energy system. For making sustainable manufacturing system, it should pass through an effective energy and a reduced carbon footprint environment, the optimum profit, and the best social effect whereas for making a reliable manufacturing system, i the manufacturing system should consistently produce quality items without any defective products. Hence, the combination of these two systems gives a sustainable economic energy efficient reliable manufacturing system. Therefore, the aim of the model is three-fold: make the manufacturing system energy efficient sustainable, continue with system reliability, and obtain the optimum profit. For making a sustainable energy efficient manufacturing system, a basic economic manufacturing quantity model is taken where the manufacturing cost is dependent on system development cost under the energy and carbon footprint consideration and production rate. The development cost (DC) under the efficient energy and carbon footprint of the manufacturing system is as follows:

$$DC = A + A' + Be^{\frac{k(\eta_{max}-\eta)}{(\eta-\eta_{min})}},\tag{2}$$

where η is the system design variable, which is an indicator of system reliability. The development cost is the summation of initial development cost, energy and carbon footprint cost, and reliability related cost.

Using this development cost per product, the unit production cost (UPC) under the optimum energy and carbon footprint can be found as the summation of material cost, energy and carbon footprint cost, development cost per product, and tool/die cost, where $S_i(t)$ is the time-dependent production rate.

$$UPC = C_{0_i} + C'_{0_i} + \frac{A + A' + Be^{\frac{k(\eta_{max} - \eta)}{(\eta - \eta_{min})}}}{S_i(t)} + \alpha S_i(t). \tag{3}$$

Therefore, the total production cost (TPC) under the effective energy per cycle is given by

$$\begin{aligned} TPC \quad &= \quad \int_0^T S_i(t) UPC dt \\ &= \quad \int_0^T \left[C_{0_i} + C'_{0_i} + \frac{A + A' + Be^{\frac{k(\eta_{max} - \eta)}{(\eta - \eta_{min})}}}{S_i(t)} + \alpha S_i(t) \right] S_i(t) dt. \end{aligned} \tag{4}$$

Therefore, for the sustainable energy efficient manufacturing system, the governing differential equation of inventory can be written as

$$\frac{dQ_i(t)}{dt} = S_i(t) - D_i(t), \tag{5}$$

where the rate of change of the production quantity is equal to the difference between time-dependent production rate and time-dependent demand.

Therefore, the production rate $S_i(t)$ can be found from Equation (5) as

$$S_i(t) = \frac{dQ_i(t)}{dt} + D_i(t). \tag{6}$$

As the product is holding for a long time by using the optimum energy carbon footprint, it is found from the old data, that the holding cost is linearly varying with time. The holding cost (HC) under the efficient energy and carbon footprint of the manufacturing system is given as follows:

$$HC = \int_0^T \left[C_{h_{1_i}} + C'_{h_{1_i}} + C_{h_{2_i}} t + C'_{h_{2_i}} t \right] Q_i dt, \tag{7}$$

where $C_{h_{1_i}}$ is the constant holding cost per unit time and $C'_{h_{1_i}}$ is the constant energy and carbon footprint cost due to holding cost; $C_{h_{2_i}}$ and $C'_{h_{2_i}}$ are similar cost but varies with time.

Due to maintaining the brand image of the manufacturing company, they arrange a full-lot inspection policy for all lots of produced products under optimum energy. Therefore, the inspection cost (IC) of the whole produced lot (Q_i), under the optimal energy and carbon footprint utilized, is given by

$$IC = \int_0^T (C_i + C'_i) Q_i dt. \tag{8}$$

To make more profit, from the previous data, it is found that the demand is dependent on time and selling price both, that is, if the time is increased then the demand is increasing exponentially, whereas if price is decreasing, then demand is increasing. Therefore, the demand is taken as

$$D_i(t, p_i) = ae^{bt} + \frac{p_{max} - p_i}{p_i - p_{min}}, \tag{9}$$

where a and b are the scaling and shape parameters for exponentially time-dependent demand,, respectively. p_i is the selling price with its maximum as p_{max} and minimum as p_{min}.

During the production system, the manufacturing system moves to an *out-of-control* state from an *in-control* state and the defective items are produced in a rate $\zeta e^{\eta t}$, where ζ is a scaling parameter.

Hence, the total defective items during time t is $\xi e^{\eta t} S_i(t)$. The reworking cost (RC) of those defective items with the minimum energy consumption and reduced carbon footprint is given by

$$RC = \int_0^T (C_{d_i} + C'_{d_i}) \xi e^{\eta t} S_i(t) dt,\tag{10}$$

where the rework cost and energy as well as carbon footprint cost per product is C_{d_i} and C'_{d_i} under the time dependent production rate $S_i(t)$.

The profit of the whole MSEEERMQ under the effective energy and carbon footprint cost is

$$\Omega_1(\eta) = \int_0^T \left[\left(p_i - \left((C_{0_i} + C'_{0_i}) + \frac{(A + A') + Be^{\frac{k(\eta_{max} - \eta)}{(\eta - \eta_{min})}}}{S_i(t)} + \alpha S_i(t) \right) \right) S_i(t) \right.$$
$$\left. - \left((C_{h_{1_i}} + C'_{h_{1_i}}) + (C_{h_{2_i}} + C'_{h_{2_i}})t \right) Q_i - (C_{d_i} + C'_{d_i}) e^{\eta t} \xi S_i(t) - (C_i + C'_i) Q_i \right] dt.\tag{11}$$

The first term of the equation, $\int_0^T p_i S_i(t) dt$, is the revenue of the MSEEERMQ system. The second term, $\int_0^T \left((C_{0_i} + C'_{0_i}) + \frac{(A + A') + Be^{\frac{k(\eta_{max} - \eta)}{(\eta - \eta_{min})}}}{S_i(t)} + \alpha S_i(t) \right) S_i(t) dt$ consists of material cost, development cost, tool/die cost along with the corresponding energy and carbon footprint costs. The third term is the constant and variable holding costs and its energy and carbon footprint costs. $\int_0^T (C_{d_i} + C'_{d_i}) e^{\eta t} \xi S_i(t) dt$ is the reworking cost along with its energy and carbon footprint cost. The last term is the inspection cost and its energy consumption cost.

Based on the previous data, it is found that the inflation rate and time-value of money are most important effective function for every cost (see for reference Govindan et al. [37]). Therefore, under inflation and time-value of money, the total profit can be calculated using Equations (6), (9) and (11)

$$\begin{aligned}
\Omega_2(\eta) \;=\; & \text{Revenue under inflation} - \text{material cost with energy and carbon footprint} \\
& \text{cost under inflation} - \text{rework cost for defective product with energy} \\
& \text{and carbon footprint cost under inflation} \\
& - \quad \text{tool/die cost with energy and carbon footprint cost under inflation} \\
& - \quad \text{holding cost with energy and carbon footprint cost under inflation} \\
& - \quad \text{inspection cost with energy and carbon footprint cost under inflation} \\
& - \quad \text{development cost with energy and carbon footprint cost under inflation} \\
\;=\; & \int_0^T e^{-\tau t} \left[\left(\dot{Q}_i + ae^{bt} + \frac{p_{max} - p_i}{p_i - p_{min}} \right) \left(p_i - (C_{0_i} + C'_{0_i}) \right. \right. \\
& \left. - (C_{d_i} + C'_{d_i}) e^{\eta t} \xi \right) - \alpha \left(\dot{Q}_i + ae^{bt} + \frac{p_{max} - p_i}{p_i - p_{min}} \right)^2 \\
& \left. - Q_i (C_{h_{1_i}} + C_{h_{2_i}} t + C_i + C'_i) - A - Be^{\frac{k(\eta_{max} - \eta)}{(\eta - \eta_{min})}} \right] dt \\
\;=\; & \int_0^T F(Q_i, \dot{Q}_i, t) dt,
\end{aligned}\tag{12}$$

where

$$F(Q_i, \dot{Q}_i, t) = e^{-\tau t}\left[\left(\dot{Q}_i + ae^{bt} + \frac{p_{max} - p_i}{p_i - p_{min}}\right)\left((p_i + p_i') - (C_{0_i} + C_{0_i}')\right)\right.$$

$$-(C_{d_i} + C_{d_i}')e^{\eta t}\zeta\Big) - \alpha\left(\dot{Q}_i + ae^{bt} + \frac{p_{max} - p_i}{p_i - p_{min}}\right)^2$$

$$\left. -Q_i(C_{h_{1_i}} + C_{h_{2_i}}t + C_i + C_i') - A - Be^{\frac{k(\eta_{max} - \eta)}{(\eta - \eta_{min})}}\right]. \tag{13}$$

To make the model sustainable, energy and carbon footprint cost are considered within the model. Hence, the final profit of the multi-item sustainable economic energy efficient reliable manufacturing quantity model under carbon footprint and energy consumption is given by

$$\Omega_3(\eta) = \int_0^T e^{-\tau t}\left[\left(\dot{Q}_i + ae^{bt} + \frac{p_{max} - p_i}{p_i - p_{min}}\right)\left(p_i - \left(C_{0_i} + C_{0_i}'\right) - \left(C_{d_i} + C_{d_i}'\right)e^{\eta t}\zeta\right)\right.$$

$$- \alpha\left(\dot{Q}_i + ae^{bt} + \frac{p_{max} - p_i}{p_i - p_{min}}\right)^2 - Q_i\left(C_{h_{1_i}} + C_{h_{1i}}' + C_{n_{2_i}}t + C_{h_{2i}}'t + C_i + C_i'\right)$$

$$\left. - (A + A') - Be^{\frac{k(\eta_{max} - \eta)}{(\eta - \eta_{min})}}\right] dt. \tag{14}$$

3.2. Solution for Optimization Problem

For the maximum profit, the simplified form of the profit can be written as

$$\Omega_3(\eta) = Y_4 + \left\{\frac{m_1 m_2(\eta + \frac{\tau}{2\alpha})}{(\eta - \tau)^2} + \frac{m_2 X_4}{\tau(\eta - \tau)}\right\}(e^{(\eta - \tau)T} - 1) - \frac{m_2 \tau N}{n}(e^{\eta T} - 1)$$

$$-\left\{\frac{m_2 a(\tau + \frac{b}{2\alpha})}{(b - \tau)(b - \tau + \eta)} + \frac{m_2 ab}{b - \tau + \eta}\right\}(e^{(b - \tau + \eta)T} - 1)$$

$$-\frac{m_2^2(\eta + \frac{\tau}{2\alpha})}{(\eta - \tau)(2\eta - \tau)}(e^{(2\eta - \tau)T} - 1) - \alpha\left[\frac{m_2^2(\eta + \frac{\tau}{2\alpha})^2}{(\eta - \tau)^2(2\eta - \tau)}(e^{(2\eta - \tau)T} - 1)\right.$$

$$\left. +\frac{2m_2 m_3(\eta + \frac{\tau}{2\alpha})}{(\eta - \tau)}(b + \eta - \tau)(e^{(b + \eta - \tau)T} - 1) - \frac{2m_2 x_4(\eta + \frac{\tau}{2\alpha})}{\tau(\eta - \tau)^2}(e^{(\eta - \tau)T} - 1)\right]$$

$$-\frac{m_7 m_2(\eta + \frac{\tau}{2\alpha})}{\eta(\eta - \tau)}(e^{(\eta - \tau)T} - 1) - \frac{m_8 m_2(\eta + \frac{\tau}{2\alpha})}{\eta(\eta - \tau)}\left\{\frac{Te^{(\eta - \tau)T}}{\eta - \tau}\right.$$

$$\left. -\frac{1}{(\eta - \tau)^2}(e^{(\eta - \tau)T} - 1)\right\} + \frac{Be^{\frac{k(\eta_{max} - \eta)}{\eta - \eta_{max}}}}{\tau}(e^{-\tau T} - 1), \tag{15}$$

(please see Appendix A for detailed calculations).

For using the necessary condition, differentiating the profit equation one can write

$$\frac{d\Omega_3(\eta)}{d\eta} = F_1(\eta) + F_2(\eta) + F_3(\eta) + F_4(\eta) + F_{5g}(\eta), \tag{16}$$

(please see Appendix B for values of $F_1(\eta)$, $F_2(\eta)$, $F_3(\eta)$, $F_4(\eta)$ and $F_{5g}(\eta)$).

The existence of the maximum value of the profit function under the effect of optimum energy through a lemma.

Lemma 1. *The profit of the whole multi-item sustainable economic energy efficient reliable manufacturing quantity model has its maximum value during the time interval $[0, T]$.*

Proof. To calculate the optimal path with optimum energy consumption and proper carbon footprint, one can assume two curves C_g and C_l, where the profit of the whole multi-item sustainable economic energy efficient reliable manufacturing quantity will be the maximum through the curve C_g and C_l is any other curve. Now, as for C_g, the profit under carbon footprint and optimum energy is maximum, thus, one can consider the curve C_g can be defined by

$$Q_i(t) = Q_{i\gamma}(t); \ t \in [0, T], \tag{17}$$

and the curve C_l can be defined as

$$Q_i(t) = Q_{il}(t) = Q_{i\gamma}(t) + lu(t); \ t \in [0, T], \tag{18}$$

where $u(t)$ is any differentiable function with respect to time t and l is very small quantity. Hence, the profit of the whole multi-item sustainable economic energy efficient reliable manufacturing quantity model under carbon footprint and optimum energy can be found by

$$\Omega_3(l) = \int_0^T v_l dt, \ \text{where } v_l = v(Q_{i\gamma}(t) + lu(t), \dot{Q}_{i\gamma}(t) + l\dot{u}(t), t). \tag{19}$$

From, the necessary condition of the classical optimization, it can be found that the rate of change of profit under carbon footprint and optimum energy with respect to l is zero as follows:

$$\left. \frac{d\Omega_3(l)}{dl} \right|_{l=0} = 0. \tag{20}$$

Therefore, from Equation (19), one can find that

$$
\begin{aligned}
\frac{d\Omega_3(l)}{dl} &= \int_0^T \left\{ u(t) \frac{\partial v_l}{\partial Q_i} + \dot{u}(t) \frac{\partial v_l}{\partial \dot{Q}_i} \right\} dt \\
&= \left[u(t) \frac{\partial v_l}{\partial \dot{Q}_i} \right]_0^T + \int_0^T u(t) \left\{ \frac{\partial v_l}{\partial Q_i} - \frac{d}{dt} \left(\frac{\partial v_l}{\partial \dot{Q}_i} \right) \right\} dt \\
&= \int_0^T u(t) \left\{ \frac{\partial v_l}{\partial Q_i} - \frac{d}{dt} \left(\frac{\partial v_l}{\partial \dot{Q}_i} \right) \right\} dt.
\end{aligned}
\tag{21}
$$

As $Q_i(t)$ is fixed at both end points during $[0, T]$, hence the arbitrary function $u(t)$ must be zero at the end points. Hence, equating $\left. \frac{d\Omega_3(l)}{dl} \right|_{l=0} = 0$ gives

$$\frac{\partial v_l}{\partial Q_i} - \frac{d}{dt} \left(\frac{\partial v_l}{\partial \dot{Q}_i} \right) = 0. \tag{22}$$

This Equation [22] represents the necessary conditions of the profit of the whole multi-item sustainable economic energy efficient reliable manufacturing quantity model. Therefore, differentiating $\frac{d\Omega_3(l)}{dl}$ with respect to l, one can find

$$\frac{d^2\Omega_3(l)}{dl^2} = \int_0^T \left\{ u(t)^2 \frac{\partial^2 v_l}{\partial Q_i{}^2} + 2u(t)\dot{u}(t) \frac{\partial^2 v_l}{\partial Q_i \dot{Q}_i} + \{\dot{u}_i\}^2 \frac{\partial^2 v_l}{\partial \dot{Q}_i{}^2} \right\} dt. \tag{23}$$

Therefore, to calculate the optimal curve, differentiating v partially with respect to $Q_i(t)$ and $\dot{Q}_i(t)$, one can find

$$\frac{\partial v}{\partial Q_i} = e^{-\tau t}\left[C_{h_{1_i}} + C'_{h_{1_i}} + C_{h_{2_i}} + C'_{h_{2_i}} + C_i + C'_i\right], \tag{24}$$

$$\frac{\partial^2 v}{\partial Q_i^2} = 0, \tag{25}$$

$$\frac{\partial v}{\partial \dot{Q}_i} = e^{-\tau t}\left[\left(p_i - C_{0_i} - C'_{0_i} - C_{d_i}e^{\eta t}\zeta - C'_{d_i}e^{\eta t}\zeta\right)\right.$$
$$\left. - 2\alpha\left(\dot{Q}_i + ae^{bt} + \frac{p_{max} - p_i}{p_i - p_{min}}\right)\right], \tag{26}$$

$$\frac{\partial^2 v}{\partial \dot{Q}_i^2} = -2\alpha e^{-\tau t}, \tag{27}$$

$$\frac{\partial^2 v}{\partial Q_i \partial \dot{Q}_i} = 0. \tag{28}$$

Now, at the point $l = 0$, one can obtain

$$\left.\frac{d^2\Omega_3(l)}{dl^2}\right|_{l=0} = \int_0^T \left\{u(t)^2\frac{\partial^2 v_l}{\partial Q_i^2} + 2u(t)\dot{u}(t)\frac{\partial^2 v_l}{\partial Q_i\partial\dot{Q}_i} + \{\dot{u}_i\}^2\frac{\partial^2 v_l}{\partial \dot{Q}_i^2}\right\}dt$$
$$= \int_0^T -2\alpha e^{-\tau t}\{\dot{u}_i\}^2 dt < 0. \tag{29}$$

Hence, this proves sufficient conditions for the whole profit of multi-item sustainable economic energy efficient reliable manufacturing quantity model under carbon footprint and optimum energy has a global maximum value within the interval $[0, T]$. This completes the proof.

For obtaining the optimum curve, from Euler-Lagrange equation, one can write

$$\frac{d}{dt}\left(\frac{\partial v}{\partial \dot{Q}_i}\right) - \frac{\partial v}{\partial Q_i} = 0 \tag{30}$$

$$\Rightarrow \frac{d}{dt}\left[e^{-\tau t}\left[\left(p_i - C_{0_i} - C'_{0_i} - C_{d_i}e^{\eta t}\zeta - C'_{d_i}e^{\eta t}\zeta\right) - 2\alpha\left(\dot{Q}_i + ae^{bt} + \frac{p_{max} - p_i}{p_i - p_{min}}\right)\right]\right]$$
$$- e^{-\tau t}\left[C_{h_{1_i}} + C'_{h_{1_i}} + C_{h_{2_i}} + C'_{h_{2_i}} + C_i + C'_i\right] = 0$$
$$\Rightarrow \ddot{Q}_i - \tau\dot{Q}_i = X_3 + X_4, \tag{31}$$

(the values of X_3 and X_4 are given in Appendix A).

By solving the Equation (31) it can be obtained

$$Q_i(t) = M_i + N_i e^{\tau t} + \frac{a\left(\tau + \frac{b}{2\alpha}\right)e^{bt}}{b(b - \tau)} + \frac{(C_{d_i} + C'_{d_i})\zeta e^{\eta t}\left(\eta + \frac{\tau}{2\alpha}\right)}{\eta(\eta - \tau)} - \frac{t}{\tau}X_4. \tag{32}$$

Using $Q_i(0) = 0$ and $Q_i(T) = 0$, it can be obtained

$$M_i = \frac{1}{(1 - e^{\tau T})} \left[\frac{TX_4}{\tau} + \frac{a\left(\tau + \frac{b}{2\alpha}\right)(e^{\tau T} - e^{bT})}{b(b - \tau)} + \frac{(C_{d_i} + C'_{d_i})\xi(e^{\tau T} - e^{\eta T})\left(\eta + \frac{\tau}{2\alpha}\right)}{\eta(\eta - \tau)} \right] \tag{33}$$

$$\text{and } N_i = \frac{1}{(e^{\tau T} - 1)} \left[\frac{TX_4}{\tau} + \frac{a\left(\tau + \frac{b}{2\alpha}\right)(1 - e^{bT})}{b(b - \tau)} + \frac{(C_{d_i} + C'_{d_i})\xi(1 - e^{\eta T})\left(\eta + \frac{\tau}{2\alpha}\right)}{\eta(\eta - \tau)} \right]. \tag{34}$$

Hence, the rate of changes of inventory and production rate can be found as

$$\dot{Q}_i(t) = \sum_{i=1}^{n} \left[\tau N_i e^{\tau t} + \frac{a\left(\tau + \frac{b}{2\alpha}\right)e^{bt}}{(b - \tau)} + \frac{(C_{d_i} + C'_{d_i})\xi e^{\eta t}\left(\eta + \frac{\tau}{2\alpha}\right)}{(\eta - \tau)} - \frac{X_4}{\tau} \right] \tag{35}$$

$$\begin{aligned} \text{and } \dot{S}_i(t) &= \sum_{i=1}^{n} \dot{Q}(t) + \sum_{i=1}^{n} D_i \\ &= \sum_{i=1}^{n} \left[\tau N_i e^{\tau t} + \frac{a\left(\tau + \frac{b}{2\alpha}\right)e^{bt}}{(b - \tau)} + \frac{(C_{d_i} + C'_{d_i})\xi e^{\eta t}\left(\eta + \frac{\tau}{2\alpha}\right)}{(\eta - \tau)} - \frac{X_4}{\tau} \right. \\ &\quad \left. + a e^{bt} + \frac{p_{max} - p_i}{p_i - p_{min}} \right]. \end{aligned} \tag{36}$$

Hence, using Equation (36), the profit becomes

$$\Omega_3(\eta) = \sum_{i=1}^{n} \int_{0}^{T} e^{-\tau t} X_8 dt - \frac{X_7(1 - e^{-\tau T})}{\tau}. \tag{37}$$

The verification of the global optimality under the optimum energy can be done by Lemma 2. $\quad\square$

3.3. Verification of Optimality Condition

The verification of optimality is proved by Lemma 2.

Lemma 2. *If the first order derivative of profit of multi-item sustainable economic energy efficient reliable manufacturing quantity model is less than zero at $\eta = \eta_{max}$ then $\frac{d\Omega_3(\eta)}{d\eta}$ must have a solution within $[\eta_{min}, \eta_{max}]$; otherwise $\frac{d\Omega_3(\eta)}{d\eta} = 0$ may or may not have an optimum solution within $[\eta_{min}, \eta_{max}]$. The maximum value of the profit of multi-item sustainable economic energy efficient reliable manufacturing quantity model exists if $\frac{d^2\Omega_3(\eta)}{d\eta^2} < 0$ at the optimum value of η^*.*

Proof. For calculation of the maximum profit of multi-item sustainable economic energy efficient reliable manufacturing quantity model, differentiating $\Omega_3(\eta)$ with respect to η two times.

Simplifying and differentiating with respect to η, one can have

$$\frac{d\Omega_3(\eta)}{d\eta} = \frac{k(1 - e^{-\tau t})}{\tau} Be^{\frac{k(\eta_{max} - \eta)}{(\eta - \eta_{min})}} \frac{(\eta_{max} - \eta_{min})}{(\eta - \eta_{min})}, \tag{38}$$

which can be further differentiated with respect to η, it can be found

$$\begin{aligned} \frac{d^2\Omega_3(\eta)}{d\eta^2} &= -\frac{k(1 - e^{-\tau t})}{\tau} Be^{\frac{k(\eta_{max} - \eta)}{(\eta - \eta_{min})}} \frac{(\eta_{max} - \eta_{min})}{(\eta - \eta_{min})^2} \left[1 + \frac{k(\eta_{max} - \eta)}{(\eta - \eta_{min})} \right] \\ &< 0. \end{aligned} \tag{39}$$

As $lim_{\eta \to \eta_{min}} \frac{d\Omega_3(\eta)}{d\eta} \to \infty$, hence $\frac{d\Omega_3(\eta)}{d\eta}$ must have at least one solution provided $\frac{d\Omega_3(\eta_{max})}{d\eta} < 0$ is satisfied; otherwise $\frac{d\Omega_3(\eta)}{d\eta} = 0$ may or may not have any optimum solution within $[\eta_{min}, \eta_{max}]$. As $\frac{d^2\Omega_3(\eta)}{d\eta^2} < 0$, hence $\Omega_3(\eta^*)$ has the maximum value at $\eta = \eta^*$ within $[\eta_{min}, \eta_{max}]$. This completes the proof. $\square$

4. Numerical Experiment

This section shows the numerical examples and sensitivity analysis of the key parameters of the study.

4.1. Numerical Example

There are three numerical examples described below.

Example 1. *For numerical experiment, the data is taken from an existing paper (Sana, [2]). For this experiment, Mathematica 10 is utilized as an optimization tool, where the computer is Intel CORE i5, 4 GB RAM. The illustrative example is given as follows:*

A MSEEERMQ model is considered for a numerical study, where the demand pattern $D_i = 150e^{2t} + \frac{p_{max} - p_i}{p_i - p_{min}}$ and cycle length (T) is 0.33 years. The management decides the maximum value of products as \$350 per product and the minimum value \$10 per product, whereas the optimum prices are decided by the management as \$50 and \$40 for two products, respectively. To hold these products for small timing in manufacturing house the constant holding costs are as \$0.8 and \$1.2 per unit product per year and the variable holding costs are \$1.5 and \$1.1 per unit product per year whereas the corresponding constant energy and carbon footprint costs are \$0.2 and \$0.3 per unit product per year as well as the variable energy consumption and carbon footprint costs are \$0.5 and \$0.4 per unit product per year. To develop two products, the basic raw material costs $(C_{0_1}$ and $C_{0_2})$ are \$2.5 per unit and \$3 per unit, respectively. The corresponding energy and carbon footprint costs to transfer those raw material into the material before using in production system are $(C'_{0_1}$ and $C'_{0_2})$ are \$0.5 per unit and \$1 per unit, respectively. The fixed development costs $(A_1$ and $A_2)$ for two products are \$4.5 and \$5.5 per product,, respectively and the corresponding energy and carbon footprint cost is $(A'_1$ and $A'_2)$ \$0.5 for each product whereas the tool/die cost (α) is \$0.1 per unit. The scaling parameter for the design variable (k) is 0.5. The cost parameter (β_i) for the system design change is \$3.5 and \$3 per product whereas the probability of the maximum value of the failure rate is 0.9 and the probability of the minimum value of the failure rate is 0.1. The defective rate of production is $\xi e^{\eta t}$, where the scaling parameter $\xi = 0.1$. The inspection costs for two products $(C_1$ and $C_2)$ are \$0.5 and \$0.4 per product. To repair those defective products, a rework cost (C_{d_i}) is utilized. Thus, the rework costs $(C_{d_1}$ and $C_{d_2})$ for two products are \$3.8 and \$0.7 per unit and energy consumption and carbon footprint costs $(C'_{d_1}$ and $C'_{d_2})$ are \$0.2 and \$0.3. For the calculation of profit, the inflation rate (τ) is considered 1% (0.01).

The global maximum profit can be found at $\eta = 0.104$ and the maximum profit is $\Omega_3(\eta) = \$2745.22/cycle$. It means that within a cycle, the number of failures is less, which is obtained for the proposed model under energy consumption and carbon footprint with the maximum profit \$2745.22 per cycle.

Example 2. *Within the data of Example 1, the value of the cycle length $T = 0.33$ time unit is changed to a $T = 0.25$ time unit; all other parameter value are the same. The optimum result is found as the maximum profit \$2043.22 per cycle and the reliability design variable $\eta = 0.17$, which means the production system is less reliable than Example 1 and less profitable. Therefore, it can be concluded that the management should maintain less cycle length to maintain the reliable sustainable production system.*

Example 3. *If the inflation rate (τ) increases from 1% to 2% and all other data is the same as Example 1, then the global maximum profit $\Omega_3(\eta) = \$2668.75$ per cycle and the failure rate $(\eta) = 0.21$. It means that with the increasing value of the inflation rate, the profit is reduced and the production system's reliability is reduced as the failure rate increases.*

4.2. Sensitivity Analysis

The sensitivity analysis of key parameters of Example 1 are considered and major findings can be concluded from the sensitivity analysis Table 3. Only some chosen parameters are written here as changes in the rest of the parameters are found negligible with a high percentage. Thus, the most sensitive parameters are taken for sensitive analysis.

Table 3. Sensitivity analysis for expected total cost.

Parameters	Changes (in %)	$\Omega_3(\eta)$ (%)	Parameters	Changes (in %)	$\Omega_3(\eta)$ (%)
$C_{h_{1_1}}$	−50%	+0.003	C_1	−50%	+0.006
	−25%	+0.002		−25%	+0.003
	+25%	−0.002		+25%	−0.003
	+50%	−0.003		+50%	−0.008
$C_{h_{1_2}}$	−50%	+0.006	C_2	−50%	0.01
	−25%	+0.003		−25%	+0.006
	+25%	−0.003		+25%	−0.007
	+50%	−0.006		+50%	−0.01
$C_{h_{2_1}}$	−50%	−0.04	P_1	−50%	−63.09
	−25%	−0.02		−25%	−30.11
	+25%	+0.02		+25%	+29.48
	+50%	+0.04		+50%	+58.65
$C_{h_{2_2}}$	−50%	−0.03	P_2	−50%	−55.79
	−25%	−0.01		−25%	−25.21
	+25%	+0.01		+25%	+31.66
	+50%	+0.03		+50%	+51.66
C_{0_1}	−50%	+3.59	C_{0_2}	−50%	+4.87
	−25%	+1.79		−25%	+2.43
	+25%	−0.86		+25%	−2.43
	+50%	−2.92		+50%	−4.87

1. With the increase of constant holding cost, the profit for both items decrease. The changes of profit are almost the same for both positive and negative changes of constant holding. Thus, it can be said that constant holding costs are less sensitive for both the items. If the energy and carbon footprint cost are increasing (which is related to the constant holding cost), the profit is decreasing too as there is no way to reduce this extra cost. Therefore, the management should take care of the amount of energy consumption even though it is less sensitive but important.

2. In case of the time-dependent holding cost, the profit is directly proportional to the cost and the corresponding energy and carbon footprint cost for both products. Thus, it can be concluded that the time-dependent holding costs show a reverse impact of constant holding cost on profit. To maintain the sustainable issue, this energy consumption and carbon footprint cost with respect to time should be maintained properly.

3. The positive changes of material costs have a negative effect on profit for both single and double items. The decreasing percentage of material cost is more effective than positive changes for single items. For both items, the percentage change of both negative and positive changes are the same. It is quite natural that the raw material's price increasing value gives less profit, but there is a consumption of energy and carbon footprint cost related to the preparation of raw material before production. If high quality raw material is utilized and the price is increased, there is a change for the reduction of energy consumption and carbon footprint for good quality raw material. Therefore, although the raw material prices increase, the profit will not be reduced by a huge amount.

4. Selling prices of two products have a huge impact on profit for both the items. The negative changes of selling prices are more sensitive than positive changes for both items. It can be found

that selling-price is most sensitive in this study. The analysis of the major key parameters are given and more analyses are not given as those are less sensitive compared to these parameters.

5. Conclusions

The study obtained a sustainable economic energy efficient reliable manufacturing system with maximum profit by using a control theory approach. The manufacturing system gave the lower amount of defective items, as the management took the reliable manufacturing system by controlling a system failure rate under the efficient use of optimum energy and carbon footprint. The demand of products for the manufacturing system was dependent on the selling-price of items. The profit is maximized globally and analytically under the efficient utilization of energy and carbon footprint at the optimum level of the failure rate, which is an indicator of system reliability. It was found that if the business production cycle was reduced, the profit was reduced but the failure rate increased, which was an indicator of a non-reliable production system through the energy consumption being obviously reduced compared to the increasing consumption due to an increased failure rate. Another major corelation was found between the inflation rate and system failure rate; the increased inflation rate gave a less reliable production system. From the sensitivity analysis, it was found that the selling-price and material cost are the most sensitive parameters compared to others. Therefore, the industry managers have to be more careful to choose perfect material and optimum selling-price to obtain a sustainable manufacturing system such that the optimum energy and carbon footprint are needed. The model considered that the time horizon was finite and that there is no shortage within the system even through defective production was there. This is the limitation of this model. Therefore, an immediate extension of this model can be considered by assuming the shortage within the model when the energy consumption and carbon footprint may take an important role. Another extension of this model can be considered as production disruption, inclusion of corrective and preventive maintenance for a multi-item production system with effectiveness of renewable and non-renewable energy. If the real data is available from an industry, then through data mining techniques, this model can be extended again for further study regarding the effect of energy in a sustainable production system. As this is a multi-item production model, the space and budget constraints can immediately be incorporated within the model, and the Kharush–Kuhn–Tucker method can be applied for optimization, which is another immediate extension of the problem.

Author Contributions: Data curation, software, formal analysis, investigation, writing—original draft preparation, M.S.; data curation—literature review, S.K.; analysis, methodology— writing, J.J.; revision, review—editing, B.G.; conceptualization, methodology, validation, visualization, funding acquisition, supervision, project administration, writing—review and editing, B.S.

Funding: The research was funded by the research fund of Hanyang University (HY-2019-N) (Project number: 201900000000419).

Acknowledgments: This work was supported by the research fund of Hanyang University (HY-2019-N) (Project number: 201900000000419).

Conflicts of Interest: The authors declare no conflict of interest.

Appendix A

$$\Omega_3(\eta) = Y_1 + \left\{ \frac{m_1 m_2 (\eta + \frac{\tau}{2\alpha})}{(\eta - \tau)^2} + \frac{m_2 X_4}{\tau(\eta - \tau)} \right\} (e^{(\eta - \tau)T} - 1) - \frac{m_2 \tau N_i}{n} (e^{\eta T} - 1)$$

$$- \left\{ \frac{m_2 a (\tau + \frac{b}{2\alpha})}{(b - \tau)(b - \tau + \eta)} + \frac{m_2 ab}{b - \tau + \eta} \right\} (e^{(b - \tau + \eta)T} - 1)$$

$$- \frac{m_2^2 (\eta + \frac{\tau}{2\alpha})}{(\eta - \tau)(2\eta - \tau)} (e^{(2\eta - \tau)T} - 1) - \alpha \left[Y_2 + \frac{m_2^2 (\eta + \frac{\tau}{2\alpha})^2}{(\eta - \tau)^2 (2\eta - \tau)} (e^{(2\eta - \tau)T} - 1) \right.$$

$$\left. + \frac{2 m_2 m_3 (\eta + \frac{\tau}{2\alpha})}{(\eta - \tau)} (b + \eta - \tau)(e^{(b + \eta - \tau)T} - 1) - \frac{2 m_2 x_4 (\eta + \frac{\tau}{2\alpha})}{\tau(\eta - \tau)^2} (e^{(\eta - \tau)T} - 1) \right]$$

$$+ Y_3 - \frac{m_7 m_2 (\eta + \frac{\tau}{2\alpha})}{\eta(\eta - \tau)} (e^{(\eta - \tau)T} - 1) - \frac{m_8 m_2 (\eta + \frac{\tau}{2\alpha})}{\eta(\eta - \tau)} \left\{ \frac{Te^{(\eta - \tau)T}}{\eta - \tau} \right.$$

$$\left. - \frac{1}{(\eta - \tau)^2} (e^{(\eta - \tau)T} - 1) \right\} + \frac{\left[(A + A') + Be^{\frac{k(\eta_{max} - \eta)}{\eta - \eta_{min}}} \right]}{\tau} (e^{-\tau T} - 1)$$

where

$$X_1 = \left[\left(p_i - C_{0_i} - C'_{0_i} - C_{d_i} e^{\eta t} \xi - C'_{d_i} e^{\eta t} \xi \right) - 2\alpha \left(\dot{Q}_i + ae^{bt} + \frac{p_{max} - p_i}{p_i - p_{min}} \right) \right]$$

$$X_2 = \left[C_{h_{1i}} + C'_{h_{1i}} + C_{h_{2i}} + C'_{h_{2i}} + C_i + C'_i \right]$$

$$X_3 = ae^{bt} \left(\tau + \frac{b}{2\alpha} \right) + (C_{d_i} - C'_{d_i}) \xi e^{\eta t} \left(\eta + \frac{\tau}{2\alpha} \right)$$

$$X_4 = -\frac{1}{2\alpha} \left[C_{h_{1i}} + C'_{h_{1i}} + C_{h_{2i}} + C'_{h_{2i}} + C_i + C'_i - 2\alpha\tau \left(\frac{p_{max} - p_i}{p_i - p_{min}} \right) + \tau(p_i - C_{0_i} - C'_{0_i}) \right]$$

$$X_5 = \left(p_i - (C_{0_i} + C'_{0_i}) - (C_{d_i} + C'_{d_i}) \xi \right)$$

$$X_6 = (C_{h_{1_i}} + C'_{h_{1_i}} + C_{h_{2_i}} t + C'_{h_{2_i}} t + C_i + C'_i)$$

$$X_7 = (A + A') + Be^{\frac{k(\eta_{max} - \eta)}{(\eta - \eta_{min})}}$$

$$X_8 = \left[N_i e^t + 2X_4 + ae^{bt} \left(\frac{b}{2\alpha} + \tau \right)(b + 1) + ae^{bt} + \frac{p_{max} - p_i}{p_i - p_{min}} \right] X_5$$

$$- \alpha \left(N_i e^t + 2X_4 + ae^{bt} \left(\frac{b}{2\alpha} + \tau \right)(b + 1) + ae^{bt} + \frac{p_{max} - p_i}{p_i - p_{min}} \right)^2$$

$$- \left[M_i + N_i e^t + 2X_4 t + ae^{bt} \left(\frac{b}{2\alpha} + \tau \right)(b + 1) \right] X_6,$$

$$
\begin{aligned}
m_1 &= (p_i - (C_{o_i} + C'_{o_i})), \\[4pt]
m_2 &= (C_{d_i} + C'_{d_i})\xi, \\[4pt]
m_3 &= \frac{a(\tau + \frac{b}{2\alpha})}{(b - \tau)}, \\[8pt]
m_4 &= \frac{p_{max} - p_i}{p_i - p_{min}}, \\[8pt]
m_5 &= \frac{a(\tau + \frac{b}{2\alpha})e^{bt}}{b(b - \tau)}, \\[8pt]
m_7 &= C_{h_{1i}} + C'_{h_{1i}} + C_i + C'_i, \\[4pt]
m_8 &= C_{h_{2i}} + C'_{h_{2i}},
\end{aligned}
$$

and

$$
Y_4 = Y_1 - \alpha Y_2 + Y_3 + \frac{(A + A')}{\tau}(e^{\tau T} - 1),
$$

$$
\begin{aligned}
Y_1 &= m_1 \tau N_i T + m_1 \frac{a(T + \frac{b}{2\alpha})}{(b - \tau)^2}(e^{(b-\tau)T} - 1) + \frac{m_1 x_4}{\tau^2}(e^{-\tau T} - 1) \\[8pt]
&\quad + \frac{m_1 ab}{(b - \tau)}(e^{(b-\tau)T} - 1) - \frac{m_1 m_4}{\tau}(e^{-\tau T} - 1) + \frac{m_2 m_4}{\tau}(e^{-\tau T} - 1)
\end{aligned}
$$

$$
\begin{aligned}
Y_2 &= \tau N_i^2(e^{\tau T - 1}) + \frac{(m_3^2 + a^2 b^2)}{(2b - \tau)}(e^{(2b-\tau)T} - 1) - \frac{(x_4^2 + \tau^2 m_4^2)}{\tau^3}(e^{-\tau T} - 1 \\[8pt]
&\quad + \frac{2(\tau^2 N_i m_3 - x_4 ab + \tau m_4 ab)}{\tau(b - \tau)}\Big)(e^{(b-\tau)T}) + 2m_4 \tau N_i T
\end{aligned}
$$

$$
\begin{aligned}
Y_3 &= m_7\Big[\frac{M_i}{\tau}(e^{-\tau T} - 1) - \frac{x_4}{\tau}\Big\{\frac{Te^{-\tau T}}{\tau} + \frac{1}{\tau^2}(e^{-\tau T} - 1)\Big\} - N_i T - \frac{m_5}{b - \tau}(e^{(b-\tau)T} - 1)\Big] \\[8pt]
&\quad + m_8\Big[M_i\Big\{\frac{T}{\tau}(e^{-\tau T} - 1)\Big\} + \frac{N_i T^2}{2} + m_5\Big\{\frac{T}{(b - \tau)}e^{(b-\tau)T} - \frac{1}{(b - \tau)^2}(e^{(b-\tau)T} - 1)\Big\} \\[8pt]
&\quad + \frac{x_4}{\tau}\Big\{\frac{T^2}{\tau}e^{-\tau T} + \frac{2}{T}\Big(\frac{Te^{-\tau t}}{\tau} + \frac{1}{T^2}(e^{-\tau T} - 1)\Big)\Big\}\Big]
\end{aligned}
$$

Appendix B

$$
\begin{aligned}
F_1(\eta) \;=\;& \frac{Tm_1m_2(\eta+\frac{\tau}{2\alpha})}{(\eta-\tau)^2}e^{(\eta-\tau)T} + \frac{Tm_2x_4e^{(\eta-\tau)T}}{\tau(\eta-\tau)} - \frac{m_2\tau N_i}{\eta^2}(e^{\eta T}-1) \\[4pt]
&+ \frac{m_1m_2}{(\eta-\tau)^2}(e^{(\eta-\tau)T}-1) - \frac{2m_1m_2(\eta+\frac{\tau}{2\alpha})}{(\eta-\tau)^3}(e^{(\eta-\tau)T}-1) \\[4pt]
&- \frac{m_2x_4}{\tau(\eta-\tau)^2}(e^{(\eta-\tau)T}-1) - \frac{TN_i\tau m_2}{\eta}e^{\eta T} \\[4pt]
&+ \frac{m_2a(\tau+\frac{b}{2\alpha})}{(b-\tau)(b-\tau-\eta)^2}(e^{-(b-\tau-\eta)T}-1) \\[10pt]
F_2(\eta) \;=\;& \frac{m_2ab(e^{(b-\tau+\eta)}-1)}{(b-\tau+\eta)^2} - \frac{aTm_2(\tau+\frac{b}{2\alpha})}{(b-\tau)(b-\tau+\eta)}e^{(b-\tau+\eta)T} - \frac{m_2abTe^{(b-\tau+\eta)T}}{(b-\tau+\eta)} \\[4pt]
&- \frac{2Tm_2^2(\eta+\frac{\tau}{2\alpha})}{(\eta-\tau)(2\eta-\tau)}e^{(2\eta-\tau)T} - \frac{m_2^2}{(\eta-\tau)^2(2\eta-\tau)^2}(e^{(2\eta-\tau)T}-1) \\[4pt]
&+ \frac{m_2^2(\eta+\frac{\tau}{2\alpha})(4\eta-3\tau)}{(\eta-\tau)^2(2\eta-\tau)^2}(e^{(2\eta-\tau)T}-1) - \frac{2\alpha m_2^2(\eta_2+\frac{\tau}{2\alpha})(e^{(2\eta-\tau)}-1)}{(\eta-\tau)^2(2\eta-\tau)} \\[10pt]
F_3(\eta) \;=\;& \frac{2\alpha m_2^2(\eta+\frac{\tau}{1\alpha})^2(3\eta-2\tau)}{(\eta-\tau)^3(2\eta-\tau)^2}(e^{(2\eta-\tau)T}-1) - \frac{2\alpha m_2^2T(\eta+\frac{\tau}{2\alpha})^2}{(\eta-\tau)^2(2\eta-\tau)}(e^{(2\eta-\tau)}-1) \\[4pt]
&- \frac{2\alpha m_2m_3}{(\eta-\tau)^2(b+\eta-\tau)}(e^{(b+\eta-\tau)T}-1) + \frac{2\alpha m_2m_3(\eta+\frac{\tau}{2\alpha})(b+2\eta-2\tau)}{(\eta-\tau)^2(b+\eta-\tau)^2} \\[4pt]
&- \frac{2\alpha Tm_2m_3e^{(b+\eta-\tau)T}}{(\eta-\tau)(b+\eta-\tau)} + \frac{2\alpha m_2x_4}{\tau}(e^{(\eta-\tau)T}-1) - \frac{2\alpha m_2x_4(\eta+\frac{\tau}{2\alpha})}{\tau(\eta-\tau)^2}(e^{(\eta-\tau)}-1) \\[10pt]
F_4(\eta) \;=\;& \frac{2\alpha m_2x_4T(\eta+\frac{\tau}{2\alpha})}{\tau(\eta-\tau)^2}e^{(\eta-\tau)T} - \frac{m_2m_1}{\eta(\eta-\tau)^2}(e^{(\eta-\tau)T}-1) \\[4pt]
&+ \frac{m_2m_7(\eta+\frac{\tau}{2\alpha})(2\eta-\tau)}{\eta^2(\eta-\tau)^2}(e^{(\eta-\tau)T}-1) - \frac{m_2m_7T(\eta+\frac{\tau}{2\alpha})}{\eta(\eta-\tau)}e^{(\eta-\tau)T} \\[10pt]
F_5(\eta) \;=\;& -\left(\frac{m_2m_8}{\eta(\eta-\tau)} - \frac{m_2m_8(\eta+\frac{\tau}{2\alpha})(2\eta-\tau)}{\eta^2(\eta-\tau)^2}\right)\left\{\frac{Te^{(\eta-\tau)T}}{(\eta-\tau)} - \frac{1}{(\eta-\tau)^2}(e^{(\eta-\tau)T}-1)\right\} \\[4pt]
&- \frac{m_2m_8(\eta+\frac{\tau}{2\alpha})}{(\eta-\tau)}\left\{\frac{T^2e^{(\eta-\tau)T}}{(\eta-\tau)} - \frac{Te^{(\eta-\tau)T}}{(\eta-\tau)^2} - \frac{Te^{(\eta-\tau)T}}{(\eta-\tau)^4} + \frac{2(e^{(\eta-\tau)T})-1}{(\eta-\tau)^3}\right. \\[4pt]
&\left. - \frac{(B+\acute{B})}{\tau}(e^{-\tau T}-1)\frac{K(\eta_{max}-\eta_{min})}{(\eta-\eta_{min})^2}\frac{K(\eta_{max}-\eta)}{(\eta-\eta_{min})}\right\}
\end{aligned}
$$

References

1. Cárdenas-Barrón, L.E.; Sarkar, B.; Treviño-Garza, G. An improved solution to the replenishment policy for the EMQ model with rework and multiple shipments. *Appl. Math. Mod.* **2013**, *37*, 5549–5554. [CrossRef]
2. Sana, S.S. A production inventory model in an imperfect production process. *Eur. J. Oper. Res.* **2010**, *200*, 451–464. [CrossRef]
3. Mettas, A. Reliability allocation and optimization for complex systems. In *Annual Reliability and Maintainability Symposium. 2000 Proceedings. International Symposium on Product Quality and Integrity*; IEEE: Piscataway, NJ, USA, 2000; pp. 216–221. [CrossRef]
4. Cao, Q.; Schniederjans, M.J. A revised EMQ/JIT production-run model: An examination of inventory and production costs. *Int. J. Prod. Econ.* **2004**, *87*, 83–95. [CrossRef]
5. Wang, C.H. The impact of a free-repair warranty policy on EMQ model for imperfect production systems. *Comput. Oper. Res.* **2004**, *31*, 2021–2035. [CrossRef]
6. Giri, B.; Dohi, T. Computational aspects of an extended EMQ model with variable production rate. *Comput. Oper. Res.* **2005**, *32*, 3143–3161. [CrossRef]

7. Sarkar, B.; Sana, S.S.; Chaudhuri, K. Optimal reliability, production lotsize and safety stock: An economic manufacturing quantity model. *Int. J. Manag. Sci. Eng. Manag.* **2010**, *5*, 192–202. [CrossRef]

8. Sarkar, B.; Sana, S.S.; Chaudhuri, K. Optimal reliability, production lot size and safety stock in an imperfect production system. *Int. J. Math. Oper. Res.* **2010**, *2*, 467–490. [CrossRef]

9. Sarkar, B.; Sana, S.S.; Chaudhuri, K. A stock-dependent inventory model in an imperfect production process. *Int. J. Proc. Manag.* **2010**, *3*, 361–378. [CrossRef]

10. Chiu, Y.S.P.; Liu, S.C.; Chiu, C.L.; Chang, H.H. Mathematical modeling for determining the replenishment policy for EMQ model with rework and multiple shipments. *Math. Comput. Mod.* **2011**, *54*, 2165–2174. [CrossRef]

11. Sarkar, B.; Sana, S.S.; Chaudhuri, K. An imperfect production process for time varying demand with inflation and time value of money—An EMQ model. *Exp. Syst. Appl.* **2011**, *38*, 13543–13548. [CrossRef]

12. Sarkar, B. An inventory model with reliability in an imperfect production process. *Appl. Math. Comput.* **2012**, *218*, 4881–4891. [CrossRef]

13. Sarkar, B.; Mandal, P.; Sarkar, S. An EMQ model with price and time dependent demand under the effect of reliability and inflation. *Appl. Math. Comput.* **2014**, *231*, 414–421. [CrossRef]

14. Sarkar, B.; Saren, S. Product inspection policy for an imperfect production system with inspection errors and warranty cost. *Eur. J. Oper. Res.* **2016**, *248*, 263–271. [CrossRef]

15. Omair, M.; Sarkar, B.; Cárdenas-Barrón, L.E. Minimum Quantity Lubrication and Carbon Footprint: A Step towards Sustainability. *Sustainablity* **2017**, *9*, 714. [CrossRef]

16. Jaber, M.Y.; Bonney, M.; Jawad, H. Comparison between economic order/manufacture quantity and just-in-time models from a thermodynamics point of view. *Comput. Ind. Eng.* **2017**, *112*, 503–510. [CrossRef]

17. Ahmed, W.; Sarkar, B. Impact of carbon emissions in a sustainable supply chain design for a second generation biofuel. *J. Clean. Prod.* **2018**, *186*, 807–820. [CrossRef]

18. Tiwari, S.; Daryanto, Y.; Wee, H.M. Sustainable inventory management with deteriorating and imperfect quality items considering carbon emission. *J. Clean. Prod.* **2018**, *176*, 154–169. [CrossRef]

19. Sarkar, M.; Sarkar, B. Optimization of safety stock under controllable production rate and energy consumption in a smart production management. *Energies* **2019**, *12*, 2059. [CrossRef]

20. Kluczek, A. An energy-led sustainability assessment of production systems—An approach for improving energy efficiency performance. *Int. J. Prod. Econ.* **2019**, *216*, 190–203. [CrossRef]

21. Harris, T.M.; Devkota, J.P.; Khanna, V.; Eranki, P.L.; Landis, A.E. Logistic growth curve modeling of US energy production and consumption. *Renew. Sustain. Energy Rev.* **2018**, *96*, 46–57. [CrossRef]

22. Dehning, P.; Blume, S.; Dér, A.; Flick, D.; Herrmenn, C.; Thiede, S. Load profile analysis for reducing energy demands of production systems in non-production times. *Appl. Energy* **2019**, *237*, 117–130. [CrossRef]

23. Khalil, M.; Berawi, M.A.; Heryanto, R.; Rizalie, A. Waste to energy technology: The potential of sustainable biogas production from animal waste in Indonesia. *Renew. Sustain. Energy Rev.* **2019**, *105*, 323–331. [CrossRef]

24. Nižetić, S.; Djilali, N.; Papadopoulos, A.; Rodrigues, J.J.P.C. Smart technologies for promotion of energy efficiency, utilization of sustainable resources and waste management. *J. Clean. Prod.* **2019**. [CrossRef]

25. Biel, K.; Glock, C.H. On the use of waste heat in a two-stage production system with controllable production rates. *Int. J. Prod. Econ.* **2016**, *181*, 174–190. [CrossRef]

26. Sarkar, B. An EOQ model with delay in payments and stock dependent demand in the presence of imperfect production. *Appl. Math. Comput.* **2012**, *218*, 8295–8308. [CrossRef]

27. Sarkar, B.; Saren, S.; Sinha, D.; Hur, S. Effect of unequal lot sizes, variable setup cost, and carbon emission cost in a supply chain model. *Math. Probl. Eng.* **2015**, *2015*, 1–13. [CrossRef]

28. San-Jośe, L.A.; Sicilia, J.; Cárdenas-Barrón, L.E.; Gutiérrez, J.M. Optimal price and quantity under power demand pattern and non-linear holding cost. *Comput. Ind. Eng.* **2019**, *129*, 426–434. [CrossRef]

29. Sarkar, B. Supply chain coordination with variable backorder, inspections, and discount policy for fixed lifetime products. *Math. Probl. Eng.* **2016**, *2016*, 1–14. [CrossRef]

30. Shin, D.; Guchhait, R.; Sarkar, B.; Mittal, M. Controllable lead time, service level constraint, and transportation discounts in a continuous review inventory model. *RAIRO-Oper. Res.* **2016**, *50*, 921–934. [CrossRef]

31. Cárdenas-Barrón, L.E.; González-Velarde, J.L.; Treviño-Garza, G. Heuristic algorithm based on reduce and optimize approach for a selective and periodic inventory routing problem in a waste vegetable oil collection environment. *Int. J. Prod. Econ.* **2019**, *211*, 44–59. [CrossRef]

32. Manente, G.; Lazzaretto, A.; Molinari, I.; Bronzini, F. Optimization of the hydraulic performance and integration of a heat storage in the geothermal and waste-to-energy district heating system of Ferrara. *J. Clean. Prod.* **2019**, *230*, 869–887. [CrossRef]

33. Govindan, K.; Jha, P.C.; Agarwal, V.; Darbari, J.D. Environmental management partner selection for reverse supply chain collaboration: A sustainable approach. *J. Environ. Manag.* **2019**, *236*, 784–797. [CrossRef] [PubMed]

34. Sarkar, B.; Majumder, A.; Sarkar, M.; Dey, B.K.; Roy, G. Two-echelon supply chain model with manufacturing quality improvement and setup cost reduction. *J. Ind. Manag. Opt.* **2017**, *13*, 1085–1104. [CrossRef]

35. Sarkar, B.; Ullah, M.; Kim, N. Environmental and economic assessment of closed-loop supply chain with remanufacturing and returnable transport items. *Comput. Ind. Eng.* **2017**, *111*, 148–163. [CrossRef]

36. Govindan, K.; Jiménez-Parra, B.; Rubio, S.; Vicente-Molina, M.A. Marketing issues for remanufactured products. *J. Clean. Prod.* **2019**, *227*, 890–899. [CrossRef]

37. Govindan, K.; Agarwal, V.; Darbari, J.D.; Jha, P.C. An integrated decision making model for the selection of sustainable forward and reverse logistic providers. *Ann. Oper. Res.* **2019**, *273*, 607–650. [CrossRef]

MDPI

Article

Economic Analysis of an Integrated Production–Inventory System under Stochastic Production Capacity and Energy Consumption

Iqra Asghar *,†, **Biswajit Sarkar** *,† and **Sung-jun Kim**

Department of Industrial and Management Engineering, Hanyang University, Ansan Gyeonggi-do 15588, Korea
* Correspondence: iqra_ntu60@yahoo.com (I.A.); bsbiswajitsarkar@gmail.com (B.S.); Fax: +82-31-400-5959 (B.S.)
† These authors contributed equally to this work.

Received: 1 June 2019; Accepted: 14 August 2019; Published: 19 August 2019

Abstract: Expensive power cost is a significant concern in today's manufacturing world. Reduction in energy consumption is an ultimate measure towards achieving manufacturing efficiency and emissions control. In the existing literature of scheduling problems, the consumption of energy is considered uncertain under the dimensions of uncertain demand and supply. In reality, it is a random parameter that also depends on production capacity, manufacturing technology, and operational condition of the manufacturing system. As the unit production cost varies with production rate and reliability of the manufacturing system, the energy consumption of the system also varies accordingly. Therefore, this study investigated an unreliable manufacturing system under stochastic production capacities and energy consumption. A stochastic production–inventory policy is developed to optimize production quantity, production rate, and manufacturing reliability under variable energy consumption costs. As energy consumption varies in different operational states of manufacturing, we consider three specific states of power consumption, namely working, idle, and repair time, for an integrated production–maintenance model. The considered production system is subjected to stochastic failure and repair time, where productivity and manufacturing reliability is improved through additional technology investment. The robustness of the model is shown through numerical example, comparative study, and sensitivity analysis of model parameters. Several graphical illustrations are also provided to obtain meaningful managerial insights.

Keywords: stochastic production capacity; unreliable production system; energy; random failure and repair rates; controllable production rate

1. Introduction

Today's fastest growing economies are forcing manufacturers to decrease the processing time of products along with providing lower costs and better quality. To maintain a better position in such a competitive market, manufacturers are needed to be highly automated, integrated, and flexible in their production capacities. Therefore, manufacturing technology is developing day by day with improved qualities such as high flexibility, smart operations, and adaptability. However, since initial capital investment and operating costs are high for smart automation, advanced manufacturing industries are concerned with managing their manufacturing system utilization to the absolute maximum limit. Regardless of acquiring elevated levels of automation, flexibility, and high-quality products, without a reliable costing mechanism, manufacturing systems cannot stay competitive with changing economic environment. Therefore, manufacturing productivity requires a stable and reliable manufacturing system for decreasing the setup time and production lead-time. Recently, Sarkar et al. [1] proposed an analytic procedure to investigate manufacturing setup and lead time reduction strategies through additional investments in manufacturing system, where setup cost improvement for a two-stage production system is addressed in [2]. Rapid economic recovery is the ultimate desire of manufacturing organizations due to intense market competition and globalization. Moreover, maintaining the productivity of the manufacturing process is the basic criterion of a manufacturing business. In this regard, Kumar et al. [3] developed structured optimal policies for the financial aspects of a product under the imperfect production process. The study shows that a manufacturing firms should devise optimal strategies for the product development process to survive in the market. Further, costs associated with an automation policy in smart manufacturing system were discussed by Sarkar et al. [4]. In this context, the significance of energy costs for the production process and manufacturing costs is worth examination as it serves as a substantial part of production costs in manufacturing industries. For instance, in the electronics industry, the energy costs represent up to 2% of total production costs. Manufacturing industry consumes up to 35% of energy production globally [5]. Similarly, in textile and printing industries, the share of energy cost is up to 2% of total production costs. The impact of energy costs is more influential in the heavy metal industry, as it takes 35% share in production costs, whereas in oil refineries, it reaches up to 65%. In the existing literature, energy cost management has been studied for factors related to generation and consumption of energy. On the other hand, this study involved improving production–inventory management with respect to energy control variables related to production and maintenance policies. These energy control variables are related to manufacturing reliability and manufacturing capacity and are improved through technology investments. Generally, energy consumption is considered as a variable or semi-variable cost item as energy consumption can reach up to 50% or more for some manufacturing organizations. Therefore, energy cost is considered as a controllable factor with critical influential effects on the expenses of manufacturing.

From an economic perspective, the lower energy consumption leads to lower manufacturing costs. From the environmental perspective, lower consumption leads to energy saving along with cost-effective factor. It reduces carbon emissions and the need to generate extra energy on behalf of Earth resources. Reducing the energy consumption of production system is also necessary because a large part of greenhouse gas emission, from production activities, comes from energy generation. In many production systems, this percentage reaches almost 100%, as stated by Bazan et al. [6]. Energy optimization problems are studied for job-shop and flow-shop system by many authors (e.g., Giret et al. [7] and Liu et al. [8]). Energy-efficient production planning (EEPP) is the basic need of manufacturing systems. In this regard, Biel and Glock [9] provided energy pricing based production planning for sustainable manufacturing system. The study considers the renewable energy incorporation in production planning along with fixed pricing strategy. Higher energy prices and strict green legislation are big challenges for today's manufacturing industry. Liu et al. [10] developed a bi-objective optimization problem to reduce idle time energy consumption and devise energy saving strategies. A comprehensive review of energy efficient production planning was provided by Biel and Glock [9]. Economic lot sizing problem with the energy consumption aspects was investigated by Rapine et al. [11], whereas energy efficient manufacturing process chains was discussed by Mousavi et al. [12]. Recently, González-Romera et al. [13] investigated the renewable energy consumption for smart community under consumer comfort level considerations. Advance equipment reduces the energy consumption significantly when consider for scheduling problems, as stated by Mori et al. [14]. Shrouf et al. [15], Tang et al. [16] and Dai et al. [17], among others, investigated the flexible flow-shop system for energy consumption. Li et al. [18] developed a scheduling problem for a flexible manufacturing system. Recently, renewable energy problems were discussed by Calicioglu et al. [19]. Smart grid system and HV grid system for high quality energy production was discussed by Pramangioulis et al. [20]. The energy management of residential hubs for demand and storage system along with environmental concerns were investigated by Brahman et al. [21]. Routing problems for energy consumption based vehicles is another area of energy management which is significantly explored by researchers [22]. More recently, Sarkar et al. [23] investigated the electric power distribution system under online to offline (O2O) supply chain management. Bazan et al. [24]

noted that no paper considered energy consumption in inventory and supply chain models before 2014. However, recently, energy consumption received the attention of researchers in the field of supply chain and inventory management. Bazan et al. [6] investigated energy consumption for production and transportation activities related to supply chain management. Bazan et al. [25] investigated closed-loop supply chain under energy usage from the production process.

Recently, Bazan et al. [24] studied supply chain coordination under energy consumption in two-echelon supply chain model. Sarkar et al. [26] studied a smart production system under energy usage. However, none of the existing models consider energy usage with an unreliable production system with stochastic machine failure and repair time. In such systems, the considerations of energy are more important compared to others, because machine breakdown increases the cost of energy usage for production process. According to Devoldere et al. [27], a certain fraction of energy consumption is always fixed for the manufacturing system irrespective of its working status. During a long-run production, a common phenomenon is the break down of the machine and production system. The system and machine break down results in higher energy usage and production of defective items [26]. The study shows that energy consumption is a critical factor as it raises the cost of the system, even if the smart production system is considered. Therefore, the consideration of energy in an unreliable production system is extremely important. However, unfortunately, energy cost receives very little attention of the researchers in the field of unreliable inventory and supply chain management. Sarkar et al. [28] studied financial implications of a random time of machine failure for a manufacturing process and developed an optimal Economic quantity model (EMQ) model with safety stock quantity. This study was further extended to imperfect production process by Sarkar et al. [29]. The study assumes that, with increased manufacturing unreliability, the manufacturing system moves from perfect production to imperfection production state. The study is further discussed with technology investment to improve the design variable of manufacturing system and reduce machine break downs [30]. Recently, Bhuniya et al. [31] extended the case of manufacturing reliability [26] for imperfect production process for bivariate dependent demand rate. Yao et al. [32] also developed manufacturing models based on manufacturing reliability for smart manufacturing. The influence of manufacturing unreliability under the effect of economical aspects is multi-fold, as investigated by Sarkar et al. [33]. The study discusses the effect of joint unreliability and inflation on an EMQ model. To increase the manufacturing reliability, this work introduces a technology development cost while treating reliability variable as a decision parameter.

Practically, the production rate of machines, intended to produce a product, can be controlled in modern manufacturing systems. If a manufacturing system allows such adjustments, production managers can directly change production rate to desired range, by changing labor working hours on manufacturing processes, and adding or removing idle hours in-between production runs. Manufacturing systems with controllable production rates are categorized as rigid or flexible production systems in the literature [34]. For rigid case, production rates are changed only at the start of a product run and then they stay constant throughout production period. In the variable/controllable scenario, the production rate is also changed during production process. Inventory literature that covers this stream of manufacturing based research focuses on production costs with respect to production rates and production up-time. Costs function associated with production rate of manufacturing system can be derived empirically, by computing the input parameter's quantity to produce one unit of finished product, at different production rates. A comprehensive overview of manufacturing literature in this research direction is given in [35]. Variable manufacturing system with unreliable machines has been discussed in the pioneering works of Rishel [36], Olsder and Suri [37] and, more recently, Glock [38]. In these studies, the machine failures and repair are discussed with Homogeneous Markov Process. For the first time, Khouja and Mehrez [39] examined a variable product rate model. They considered the production rate dependent production quality of system. Supply chain management under the joint effect of time varying holding cost and variable production rate was examined by Sarkar et al. [40]. Impact of variable production rate in a two-stage production system was studied by Alfares and Ghaithan [41]. They suggested that variable production rates also gives flexibility in production planning to smooth the flows of raw material and work-in-process inventories which helps to avoid bottlenecks at working stations. Production rate is considered as a dynamic variable in [42]. A single vendor, multi-buyer supply chain model is studied in [43], where vendor production cost changes with variable production rate. For multiple-buyers, a continuous-review inventory model is developed to inspect inventory levels in this study. In addition, a crashing cost is employed to reduce buyer's lead times. One major flaw in this research stream is that it only focuses on unit production cost associated with variable production rates. The effect of variable production rate on other production and inventory related costs such as maintenance, storage, and carrying is neglected in the existing research. In addition, the impact of energy consumption in the dimension of unreliable and random capacity manufacturing operations is still underdeveloped. The development of technology, advances the performance of manufacturing machines, and reduces production up-time by proposing a controllable production rate in manufacturing units. However, higher production rates would need higher operating/maintenance costs and may increase the machine failure rate. With controllable production rate, an increase of processing load on the manufacturing system leads to higher costs of energy consumption per unit time. Consequently, it is inevitable to change spare parts and plan maintenance schedules frequently for manufacturing system, and then the cost of maintenance would also rise. This way, manufacturing reliability and production rates become the energy control variables. Ultimately, it is important to examine the effect of manufacturing unreliability on variable production capacities and associated energy consumption. Therefore, this study investigated the impact of optimum production lot size, production rate and manufacturing reliability on manufacturing energy consumption. In particular, we optimize the integrated production–inventory–maintenance policy when the manufacturing system faces random failures and random repair times. The manufacturing energy consumption is considered accordingly for three manufacturing states of production system: in operation, in idle state and for system repair work.

In this paper, energy consumption with respect to consumed electric power for each manufacturing task and energy control variables is considered. This paper investigates the variable energy consumption effect with controllable production rate and gives the optimal level for additional tooling and technology developmental costs. The relationship between variable energy consumption and energy control factors (production rate and failure rate) is justified on two criteria. The first is that, as the controllable production rate in units/unit time is changed, the associated specific energy consumption rate (SEC) for manufacturing process in kWh/units will change accordingly. The higher production rate reduces the specific energy consumed per item for manufacturing process. With the reduction of production time, produced quantity counterbalances for the rise of energy consumed per hour. Similarly, as the number of manufacturing failures increases, the energy consumption per units produced will increase and accordingly extra energy requirements for production restoration and start-ups will arise. Increased number of manufacturing system break downs lead to increased machine warming and restore/re-setup energy, and thus it increases the energy consumption per unit of production. This way, variable energy consumption for manufacturing process has three different components: energy consumption during setup and operational time, energy consumption during idle time due to break down and energy requirements for maintenance/re-setup activities of manufacturing system. In this paper, production, and maintenance policy related energy control factors are improved to optimize energy consumption in the manufacturing industry. Since machine failure rate is highly dependent on the design variable of manufacturing system, quality of maintenance, and production rates, this paper investigates the electrical energy consumption rates of an unreliable manufacturing system with respect to maintenance and production management factors, namely production rate p and failure rate ϕ. Table 1 provides the existing contribution efforts of different authors in the proposed research area.

Table 1. Research contributions by different authors.

Author(s)	Manufacturing Reliability	Maintenance Policy	Production Capacity	Development Cost	Energy
Marchi et al. [44]	variable	NA	variable	NA	considered
Marchi et al. [45]	variable	NA	variable	NA	considered
Bhuniya et al. [31]	variable	NA	variable	(ϕ)	considered
Sarkar et al. [26]	variable	NA	variable	(ϕ)	considered
Adane and Nicolescu [46]	NA	considered	constant	NA	NA
Demichela et al. [47]	NA	considered	NA	NA	considered
Li et al. [18]	NA	NA	flexible	NA	NA
Rackow et al. [48]	NA	NA	energy flexible	NA	considered
Shibin et al. [5]	NA	NA	flexible	NA	considered
Lee and Prabhu [49]	NA	considered	energy-aware	NA	considered
This study	stochastic	considered	controllable	(p, ϕ)	considered

(ϕ) means reliability dependent development cost. (p, ϕ) means reliability and production capacity dependent development cost. NA means not-applicable.

The rest of this paper is structured as follows. Section 2 addresses mathematical model formulation. Section 3 develops mathematical modeling of the proposed unreliable manufacturing system. Section 4 illustrates the developed mathematical model with numerical experiment and sensitivity analysis of model parameters. Finally, Section 5 concludes the study and suggests some future extensions.

2. Problem Statement, Notation, and Assumptions

This sections defines the problem statement and assumptions for the proposed mathematical model and notation used to develop mathematical model, which are given below.

2.1. Problem Statement

In this paper, an integrated production–inventory–maintenance policy is optimized under energy control variables named as production rate and manufacturing reliability. Moreover, the specific production energy consumption for each product is considered as a component of per unit manufacturing cost. The paper considers an unreliable manufacturing system which can be regulated at different production rates within design limits ($p \in [p_{min}, p_{max}]$), resulting in different unit production costs. The variable production rate decrease production lead-time and per unit production cost. On the contrary, other manufacturing operation costs related to maintenance and spare parts replacement raise with higher production rates. The manufacturing system is subjected to stochastic failure rate and, on each random breakdown, it is immediately subjected to corrective repair. A minor restoration is required along with each corrective repair, which is also stochastic in nature. Now, optimal integrated production planning is developed under the influence of energy control variables, where, with the increase in production rate, the production up-time is reduced which also reduces the energy consumption for manufacturing. On the other side, higher production rates leads to higher break downs of system and energy requirements for maintenance activities. To increase productivity of manufacturing system, a production-technology development cost is considered. Moreover, to increase the reliability of manufacturing system, a separate technology development cost is also added in the system costs. Based on different operation states of manufacturing process, energy costs for three states of manufacturing system are considered: working state (setup + production), idle time (break down + process disruptions), and for maintenance activities (repair + restoration).

Notation

Decision variables

Q	Production lot size (units)
p	Production rate, where $p \in [p_{min}, p_{max}]$ (units/unit time)
ϕ	The manufacturing reliability parameter, where $\phi \in [\phi_{min}, \phi_{max}]$

Model parameters

T	Replenishment cycle time (year)
S_m	Setup cost (\$/setup)
J	Restoration cost (\$/setup)
x	Demand rate (units/unit time)
$C(p, \phi)$	Unit production cost, function of reliability and productivity (\$)
M	Raw material cost (\$/unit)
r	Raw material increment cost due to manufacturing unreliability (\$/unit)
L	Unit manufacturing yield loss cost due to manufacturing unreliability (\$/unit/unit time)
ω	Tool/die cost (\$/unit)
γ	Scale parameter
A	Production-technology development cost, cost to increase production rate by one unit (\$)
α	Productivity scale parameter
β	Productivity shape parameter
R_m	Technology development cost (\$)
ζ	Reliability constant, non-negative
ϱ	Reliability scale parameter
C_h	Inventory holding cost (\$/unit/unit time)
σ	Repair rate (repairs/unit time)
C_r	Corrective repair cost (\$/unit time)
C_s	Opportunity cost due to lost sales (\$/lost unit)
(t_1)	Random variable denoting time to manufacturing system failure (non-negative)
(t_2)	Random variable denoting time to repair upon system failure (non-negative)
$f(t_1), F(t_1)$	Probability density function, cumulative density function of t_1
$f(t_2), F(t_2)$	Probability density function, cumulative density function of t_2
TC_1	Expected total cost per cycle with manufacturing system failure (\$)
TC_2	Expected total cost per cycle without manufacturing system failure (\$)
T_1	Expected cycle length with manufacturing system failure (time)
T_2	Expected cycle length without manufacturing system failure (time)
ETC_1	Expected total cost per unit time with manufacturing system failure (\$)
ETC_2	Expected total cost per unit time without manufacturing system failure (\$)
ETC	Expected total cost per unit time of manufacturing system (\$)

2.2. Assumptions

To model the proposed scenario mathematically, the following assumptions, apart from the previously stated ones, are considered.

- The demand rate of the products is known and constant, whereas the controllable production rate of manufacturing system is greater than demand rate $(p > x)$, and a decision variable.
- The unit production cost is not constant but a function of controllable processing rate and manufacturing unreliability.
- The reliability of the manufacturing system is derived through a stochastic process.
- The reliability of manufacturing system is expressed as a design variable, where it is represented as $\phi = \frac{Total\ number\ of\ failure}{Total\ working\ hours}$, and reliability can be increased through investment in production technology and manufacturing resources.
- The failure rate and repair rate are random variables and follow continuous probability distributions.
- The manufacturing reliability and production rate are considered as energy control variables, independent and additive.
- Shortages are considered for longer repair duration.

3. Mathematical Model

This paper investigates an unreliable production–inventory model under controllable production rate and variable electrical energy consumption effects. In this context, an integrated production–maintenance policy is designed under the influence of random failure rate and random repair rate. The manufacturing system starts at time $(t = 0)$ with a controllable production rate $(p \in [p_{min}, p_{max}])$ to manufacture a production quantity Q. The production proceed until production up-time $\frac{Q}{p}$ and the inventory accumulates with the rate $(p - x)t_m$, if manufacturing system does not face any failure. The on-hand inventory is driven through differential equation as:

$$\frac{dQ_1(t)}{dt} = (p - x), \qquad (0 \leq t \leq t_m), \quad \text{with the initial condition } Q_1(0) = 0,$$

and

$$\frac{dQ_2(t)}{dt} = (-x), \qquad (t_m \leq t \leq T), \quad \text{with the given condition } I_1(t_m) = I_2(t_m).$$

The solution of above equations result,

$$I(t) = \begin{bmatrix} (p - x)t & 0 \leq t \leq t_m \\ pt_m - xt & t_m \leq t \leq T. \end{bmatrix}$$

Since manufacturing system may face random breakdowns during production up-time, two cases arise for production planning. If time to breakdown is $(t < t_m)$, then a random breakdown take place during production up-time t_m and corrective maintenance is started immediately. If failure occurs at $(t \geq t_m)$, then no failure is considered during production up-time. Therefore, two separate models are examined in the following section, as shown in Figure 1.

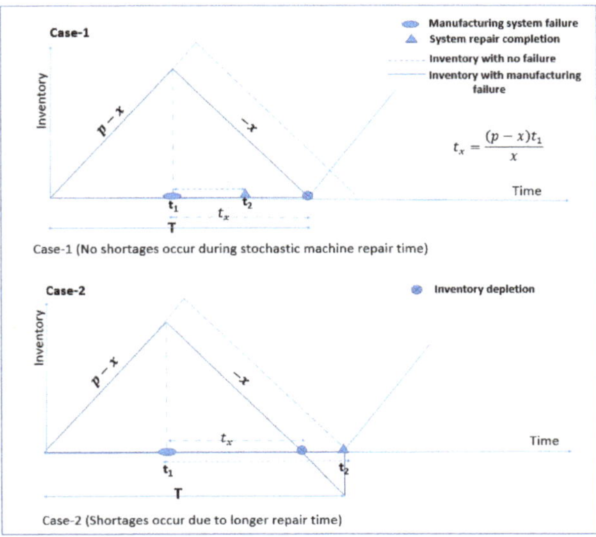

Figure 1. Inventory diagram for an unreliable manufacturing system.

3.1. Model 1: Random Production Capacity Model with Manufacturing System Failure

In this case, at time $(t = t_1)$, the manufacturing system deteriorates and production system stops. The inventory accumulates during time interval $[0, t_1]$ and depletes with demand rate x during $[t_1, T]$. Therefore, the expected total inventory becomes:

$$\int_0^{t_1} (p - x)t\, dt + \int_{t_1}^T (pt_1 - xt)\, dt.$$

As the repair rate is also a random variable, during stochastic repair time, two further cases may occur, as described in Figure 1.

Case 1 (no shortages occur during stochastic repair time)

This scenario is depicted in Case 1 of Figure 1, where the required time to repair manufacturing system after occurrence of failure is small enough that it lies within the on-hand inventory depletion time $\frac{(p-x)t_1}{x}$; then, expected inventory becomes,

$$\text{Inv}_1 = \int_0^{\frac{Q}{p}} \left(\int_0^{\frac{(p-x)t_1}{x}} \left(\frac{p(p-x)t_1^2}{2x} \right) f(t_2)\, dt_2 \right) f(t_1)\, dt_1.$$

Case 2 (shortages occur during stochastic repair time)

Case 2 of Figure 1 represents the scenario of stochastic repair time where the required time to repair manufacturing system $(t = t_2)$ after failure is large enough that it exceeds the on-hand inventory depletion time $\frac{(p - x)t_1}{x}$; then, expected inventory becomes,

$$\text{Inv}_2 = \int_0^{\frac{Q}{p}} \left(\int_{\frac{(p-x)t_1}{x}}^{\infty} \left(\frac{p(p-x)t_1^2)}{2x} \right) f(t_2)\, dt_2 \right) f(t_1)\, dt_1,$$

and the expected shortages due to longer repair time are given as,

$$\int_0^{\frac{Q}{p}} \left(\int_{\frac{(p-x)t_1}{x}}^{\infty} (xt_2 - (p - x)t_1) f(t_2)\, dt_2 \right) f(t_1)\, dt_1.$$

Total production–maintenance cost per cycle TC_1 for combining the two cases consists of the variable setup cost, variable restoration cost, variable production-technology development cost, variable production costs, variable cost for repairing the manufacturing system, inventory holding costs, and shortage cost, which are given below.

Energy costs

The cost of energy consumption is considered as a variable/semi-variable cost parameter in the manufacturing industry. The management of energy variables is very important as the cost of energy consumption can reach up to 50% or more of the variable costs category of manufacturing costs. Therefore, to include energy costs for the unreliable manufacturing system, several new notations are introduced as follows.

Energy parameters

C_{elc}	Unit cost of electrical energy (\$/kWh)
E_p	Expected energy cost for production (\$)
e_1	Fixed component of required energy for manufacturing system (kWh/time)
e_2	Variable component of required energy for production process (kWh/unit)
E_r	Expected energy cost for machine repair (\$)
e_r	Energy required for repair time (kWh/unit time)
e_s	Energy required for each production setup (kWh/setup)
e_{rs}	Energy required for each minor restoration (kWh/setup)
e_h	Energy required for inventory holding (kWh/unit/unit time)

Marchi et al. [45] referred to production energy consumption as proportional to manufacturing rate, therefore, under variable production rate, a manufacturing system develops variable energy consumption states. As stated by Gutowski et al. [50], the average energy cost for production E_p depends on the specific energy consumption per unit produced (SEC) which consist of two components: constant energy requirement e_1 (kWh/time) and variable energy requirement e_2 (kWh/unit) for production process. Energy component e_1 is required for ready state functions of manufacturing system such as (e.g., lubrication, cooling, humidification, pressure pumps, and centrifugal energy) and variable energy component e_2 depends on the production rate of manufacturing system, therefore SEC for production becomes (see Bazan et al. [6]):

$$SEC(p) = \frac{e_1}{p} + e_2.$$

Energy costs for three states of manufacturing system for a production run are considered in the following section: working state (setup + production time), idle time (break down + process disruptions), and for maintenance activities (repair + restoration).

Setup cost

Setup cost for a manufacturing system varies with energy consumption for each production quantity and production rate. Frequent failures of manufacturing systems lead to raised warming of the system and consequent restart-up energy, which directly increases the energy consumption per unit of production. Considering that, generally, the failure rate of the manufacturing system is profoundly dependent on the quality of the operating system, this paper considers restoration cost as a function of reliability and initial setup cost of manufacturing system. In addition, as the manufacturing system faces random failure rate, on each failure ϕ, a minor setup is carried out to restore the manufacturing system to initial operational conditions at a cost ($\frac{IS_m}{\phi}$). Thus, the model considers variable setup cost, variable restoration cost, expected energy consumption cost per setup, and expected energy cost for restoration for the manufacturing system as follow,

$$SC(\phi) = S_m \left(1 + \frac{J}{\phi} \right) + C_{elc}e_s + \frac{C_{elc}e_{rs}}{\phi}. \tag{1}$$

Production cost

Productivity of an automated manufacturing system varies with the operating conditions, manufacturing system process quality and manufacturing costs. With variable productivity, the energy consumption of manufacturing system also varies. The variable manufacturing rate reduces the production up-time and production lead time, however some operational conditions such as cost of processing tools, maintenance and energy consumption increase conversely. As the manufacturing reliability defines the productivity per unit of time for manufacturing system, the productivity itself is an operational concept which shows the manufacturing throughput per unit of time. To improve the energy consumption and productivity of the manufacturing system, we consider a controllable production rate, which varies within designed limits $[p_{min}, p_{max}]$. As the production rate is a variable parameter, the unit production cost depends on energy control variables, production rate and failure rate of manufacturing system, as given below,

$$C(p, \phi) = \frac{(c_p + L\phi)}{p} + p^\gamma \omega + (M + r\phi). \tag{2}$$

The unit production cost consists on several production rate and manufacturing reliability based components, which are explained below:

1. The cost component $(M + r\phi)$ is the cost of raw material, which increases linearly with increasing failure rate of manufacturing system.
2. The c_p component represents the manufacturing costs, i.e., capital and labor costs, and, as the production rate increases, manufacturing cost is evenly distributed across a vast number of produced units. Consequently, the per unit manufacturing cost reduces with higher production rates (see Figure 2).
3. The next cost component is $L\phi$, which represents the manufacturing yield loss. As the failure rate of manufacturing system increases, the downtime increases and overall costs of system increases linearly (see Figure 3). However, with higher production rates, the unit yield loss cost decreases as the higher production rate reduces the production up-time and more items are produced in fewer working hours.
4. The last cost component is referred to as tool/die cost, and it is directly proportional to production rate.
5. The energy cost for production E_p is given as a function of electrical energy consumption C_{elec} (kWh) based on controllable production rate. Therefore, expected energy cost for a production run is derived as a function of unit electricity cost C_{elc} ($/kWh), specific energy required per item produced $SEC(p)$ (kWh/unit), production up-time t_m (hours), and production rate p (units/unit time).

$$E_p \left(C_{elec} \right) = C_{elc} SEC(p) p t_m.$$

Moreover, as the unit manufacturing cost is equally distributed over a wide range of produced products, per unit production cost becomes,

$$C(p, \phi) = \frac{(c_p + e_1 + L\phi)}{p} + e_2 + p^\gamma \omega + (M + r\phi). \tag{3}$$

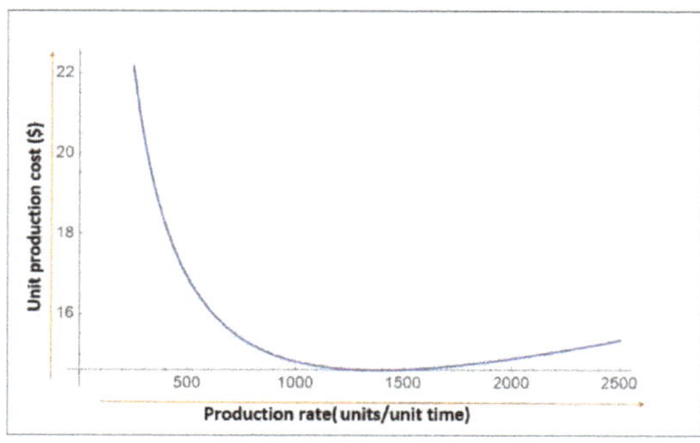

Figure 2. Change in per unit manufacturing cost $C(p, \phi)$ with varying production rate (p).

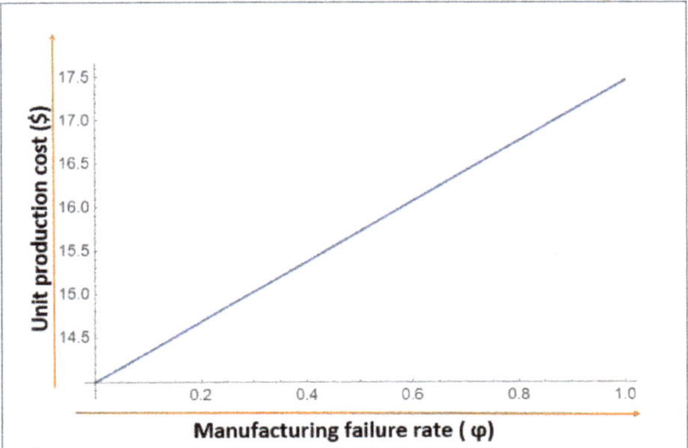

Figure 3. Change in per unit manufacturing cost $C(p,\phi)$ with varying manufacturing reliability (ϕ).

As illustrated in Figure 4, specific energy consumption per unit $SEC(p)$ decreases with an increase in production rate for the proposed model and eventually energy required to produce a single unit decreases [51].

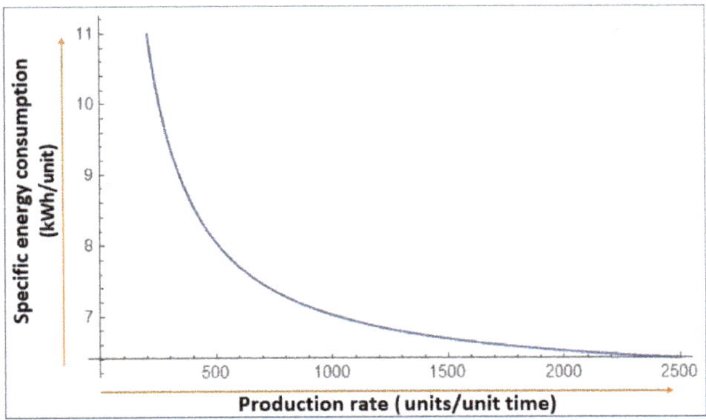

Figure 4. Change in specific energy cost (SEC) per unit manufactured with varying production rate (p).

Hence, the expected total production cost becomes,

$$CP_1(p,\phi) = \int_0^{\frac{Q}{p}} \left(\int_0^{\frac{(p-x)t_1}{x}} \left(pt_1 \left(\frac{c_p + C_{elc}e_1 + L\phi}{p} + C_{elc}e_2 + (M + r\phi) + p^\gamma \omega \right) \right) f(t_2)\, dt_2 \right) f(t_1)\, dt_1$$

$$+ \int_0^{\frac{Q}{p}} \left(\int_{\frac{(p-x)t_1}{x}}^{\infty} \left(pt_1 \left(\frac{c_p + C_{elc}e_1 + L\phi}{p} + C_{elc}e_2 + (M + r\phi) + p^\gamma \omega \right) \right) f(t_2)\, dt_2 \right) f(t_1)\, dt_1.$$

As the production rate is measured as units/unit time, and the energy consumption is derived in kWh/units of production, which infer that, as the production rate increases, the specific energy consumption per unit manufactured decreases for given production up-time. In addition, higher production rates lead to higher utilization of machine working time and more goods are produced in less time. The same assertions are made in other recent studies [6,44,45].

Production-technology development cost

The productivity of a manufacturing system depends on a combination of system design parameters named as production capacity and production reliability. In modern automated manufacturing system, the manufacturing reliability and productivity can be improved through investment in advanced production technology, where manufacturing systems with more advanced technology are considered more reliable and efficient. For controllable production rates, the increase in the processing load of the production system increases the consumed electric power per unit time. As the manufacturing system can operate at higher production rates ($p \in [p_{min}, p_{max}]$), the working life of manufacturing components shorten over time and the required frequency of tool change increases. Therefore, manufacturing systems with controllable production rates exhibit higher costs of processing tools for sustaining higher productivity. Eventually, it is essential to replace spare parts/tools of the manufacturing system quite often, which then increases the cost of manufacturing tools. To yield the development of manufacturing system to sustain higher productivity, we consider production-technology development cost $G(p, \phi)$, as illustrated in Figure 5, as a function of controllable production rate and variable manufacturing reliability. Figure 5 shows that an increase in production unit by one unit results in the elevated machine stress requiring extra management costs as given below,

$$G(p, \phi) = A\alpha \left(\frac{p - p_{min}}{p_{max} - p} \right)^{\beta} + R_m e^{\frac{\zeta\phi}{\varrho}}, \tag{4}$$

where A is the cost of labor. Energy and extra management costs, which are dependent on manufacturing rate, α and β, are non-negative production constants. The productivity of the manufacturing system also depends on the operational condition of system, which is named as manufacturing reliability. To improve manufacturing reliability within the achievable limits, ($\phi \in [\phi_{min}, \phi_{max}]$), a variable technology development cost is considered, as shown in Figure 6. R_m is the cost of technology, resources, and energy for manufacturing system when it is absolutely reliable with failure rate ($\phi = \phi_{min}$). The constant ζ represents the difficulties in increasing reliability parameter due to limited technological resources and manufacturing system complexities, and ϱ is a scale parameter, which shows the effectiveness of advanced manufacturing technology.

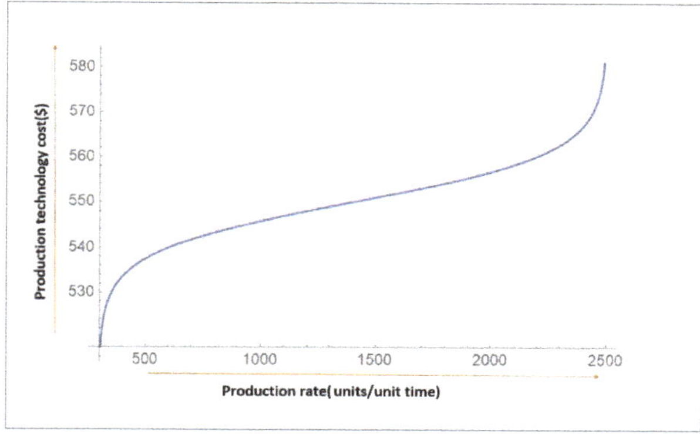

Figure 5. Change in production-technology development cost $G(p, \phi)$ with varying production rate (p).

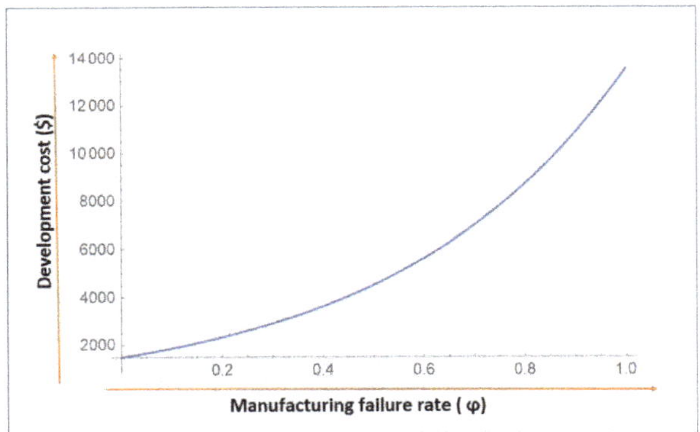

Figure 6. Change in production-technology development cost $G(p, \phi)$ with varying manufacturing reliability (ϕ).

Inventory holding cost

The expected inventory holding cost for Model 1 under the energy control variables is a sum of carrying cost and required energy cost for inventory, given as,

$$IC = (C_h + C_{elc}e_h) \int_0^{\frac{Q}{p}} \left(\int_0^{\frac{(p-x)t_1}{x}} \left(\frac{pt_1^2(p-x)}{2x} \right) f(t_2) \, dt_2 \right) f(t_1) \, dt_1$$

$$+ (C_h + C_{elc}e_h) \int_0^{\frac{Q}{p}} \left(\int_{\frac{(p-x)t_1}{x}}^{\infty} \left(\frac{pt_1^2(p-x)}{2x} \right) f(t_2) \, dt_2 \right) f(t_1) \, dt_1.$$

Corrective repair cost

On each failure, the manufacturing system is subjected to immediate corrective repair, and abort/resume (AR) inventory policy is applied. During stochastic repair time t_2, the expected energy consumption cost E_r is given as,

$$E_r(C_{elec}) = C_{elc}e_r t_2.$$

Hence, the expected repair cost is given as,

$$RC = (C_r + C_{elc}e_r) \int_0^{\frac{Q}{p}} \left(\int_0^{\infty} t_2 f(t_2) \, dt_2 \right) f(t_1) \, dt_1,$$

Therefore, expected total cost for Model 1 becomes,

$$
ETC_1(Q, p, \phi) = S_m\left(1 + \frac{J}{\phi}\right) + C_{elc}e_s + \frac{C_{elc}e_{rs}}{\phi} + A\alpha\left(\frac{p - p_{min}}{p_{max} - p}\right)\beta + R_m e^{\frac{\iota\phi}{\varrho}} + \int_0^{\frac{Q}{p}}\left(\int_0^{\frac{(p-x)t_1}{x}}\right.
$$

$$
\left(pt_1\left(\frac{c_p + C_{elc}e_1 + L\phi}{p} + C_{elc}e_2 + (M + r\phi) + p^\gamma\omega\right)\right) f(t_2)\,dt_2\right) f(t_1)\,dt_1 + \int_0^{\frac{Q}{p}}\left(\int_{\frac{(p-x)t_1}{x}}^{\infty}\right.
$$

$$
\left(pt_1\left(\frac{c_p + C_{elc}e_1 + L\phi}{p} + C_{elc}e_2 + (M + r\phi) + p^\gamma\omega\right)\right) f(t_2)\,dt_2\right) f(t_1)\,dt_1 + (C_h + C_{elc}e_h)\int_0^{\frac{Q}{p}}
$$

$$
\left(\int_0^{\frac{(p-x)t_1}{x}}\left(\frac{pt_1^2(p - x)}{2x}\right) f(t_2)\,dt_2\right) f(t_1)\,dt_1 + (C_h + C_{elc}e_h)\int_0^{\frac{Q}{p}}\left(\int_{\frac{(p-x)t_1}{x}}^{\infty}\left(\frac{pt_1^2(p - x)}{2x}\right)\right.
$$

$$
\left. f(t_2)\,dt_2\right) f(t_1)\,dt_1 + (C_r + C_{elc}e_r)\int_0^{\frac{Q}{p}}\left(\int_0^{\infty} t_2 f(t_2)\,dt_2\right) f(t_1)\,dt_1 + C_s\int_0^{\frac{Q}{p}}\left(\int_{\frac{(p-x)t_1}{x}}^{\infty}(xt_2 -\right.
$$

$$
\left.\left.(p - x)t_1\right) f(t_2)\,dt_2\right) f(t_1)\,dt_1. \tag{5}
$$

Similarly, the expected cycle length for Model 1 is defined as the time interval between two production runs with conditioning on the time to failure and time to repair:

$$
T_1(Q, p, \phi) = \int_0^{\frac{Q}{p}}\left(\int_0^{\frac{(p-x)t_1}{x}}\left(\frac{pt_1}{x}\right) f(t_2)\,dt_2 + \int_{\frac{(p-x)t_1}{x}}^{\infty}(t_1 + t_2) f(t_2)\,dt_2\right) f(t_1)\,dt_1. \tag{6}
$$

3.2. Model 2: Random Production Capacity Model without System Failure

When the time for stochastic failure is greater than the planned production up-time as $(t \geq t_m)$, then no break down is considered during production run. Hence, the expected total production cost becomes,

$$
CP_2(p, \phi) = \int_{\frac{Q}{p}}^{\infty}\left(Q\left(\frac{c_p + C_{elc}e_1 + L\phi}{p} + C_{elc}e_2 + (M + r\phi) + p^\gamma\omega\right)\right) f(t_1)\,dt_1.
$$

Therefore, the total costs of the system for TC_2 becomes setup cost, variable production-technology cost, variable production cost, variable energy consumption cost and variable inventory holding cost, given as,

$$
\begin{aligned}
ETC_2(Q, p, \phi) &= S_m + C_{elc}e_s + A\alpha\left(\frac{p - p_{min}}{p_{max} - p}\right)\beta + (C_h + C_{elc}e_h)\int_{\frac{Q}{p}}^{\infty}\left(\frac{Q^2(p - x)}{2px}\right) f(t_1)\,dt_1 \\
&+ \int_{\frac{Q}{p}}^{\infty}\left(Q\left(\frac{c_p + C_{elc}e_1 + L\phi}{p} + C_{elc}e_2 + (M + r\phi) + p^\gamma\omega\right)\right) f(t_1)\,dt_1.
\end{aligned} \tag{7}
$$

Moreover, the expected cycle length of production–inventory cycle for Model 2 is obtained as,

$$
T_2(Q, p, \phi) = \int_{\frac{Q}{p}}^{\infty}\left(\frac{Q}{x}\right) f(t_1)\,dt_1. \tag{8}
$$

3.3. Integrated Stochastic Capacity-Reliability Model

The expected total cost of integrated production–inventory model with/without stochastic failures is the sum of Models 1 and 2, which is given as,

$$
ETC(Q, p, \phi) = S_m\left(1 + \frac{J}{\phi}\right) + C_{elc}e_s + \frac{C_{elc}e_{rs}}{\phi} + A\alpha\left(\frac{p - p_{min}}{p_{max} - p}\right)^\beta + R_m e^{\frac{\zeta\phi}{\varrho}} + \int_0^{\frac{Q}{p}}\left(\int_0^{\frac{(p-x)t_1}{x}}\right.
$$

$$
\left(pt_1\left(\frac{c_p + C_{elc}e_1 + L\phi}{p} + C_{elc}e_2 + (M + r\phi) + p^\gamma\omega\right)\right)f\left(t_2\right)dt_2\right)f\left(t_1\right)dt_1 + \int_0^{\frac{Q}{p}}\left(\int_{\frac{(p-x)t_1}{x}}^\infty\right.
$$

$$
\left(pt_1\left(\frac{c_p + C_{elc}e_1 + L\phi}{p} + C_{elc}e_2 + (M + r\phi) + p^\gamma\omega\right)\right)f(t_2)\,dt_2\right)f(t_1)\,dt_1 + \int_{\frac{Q}{p}}^\infty\left(Q\left(C_{elc}e_2\right.\right.
$$

$$
\left.\left.+\frac{c_p + C_{elc}e_1 + L\phi}{p} + (M + r\phi) + p^\gamma\omega\right)\right)f(t_1)\,dt_1 + (C_h + C_{elc}e_h)\int_0^{\frac{Q}{p}}\left(\int_0^{\frac{(p-x)t_1}{x}}\left(\frac{pt_1^2(p-x)}{2x}\right)\right.
$$

$$
\left.f(t_2)\,dt_2\right)f(t_1)\,dt_1 + (C_h + C_{elc}e_h)\int_{\frac{Q}{p}}^\infty\left(\frac{Q^2(p-x)}{2px}\right)f(t_1)\,dt_1 + (C_h + C_{elc}e_h)\int_0^{\frac{Q}{p}}\left(\int_{\frac{(p-x)t_1}{x}}^\infty\right.
$$

$$
\left(\frac{pt_1^2(p-x)}{2x}\right)f(t_2)\,dt_2\right)f(t_1)\,dt_1 + (C_r + C_{elc}e_r)\int_0^{\frac{Q}{p}}\left(\int_0^\infty t_2 f(t_2)\,dt_2\right)f(t_1)\,dt_1 + C_s\int_0^{\frac{Q}{p}}
$$

$$
\left(\int_{\frac{(p-x)t_1}{x}}^\infty (xt_2 - (p - x)t_1)f(t_2)\,dt_2\right)f(t_1)\,dt_1. \tag{9}
$$

The expected cycle length from Equations (6) and (8) as a time interval, between two successive production runs is given as,

$$
E[T](Q, p, \phi) = \int_0^{\frac{Q}{p}}\left(\int_0^{\frac{(p-x)t_1}{x}}\left(\frac{pt_1}{x}\right)f(t_2)\,dt_2 + \int_{\frac{(p-x)t_1}{x}}^\infty (t_1 + t_2)f(t_2)\,dt_2\right)f(t_1)\,dt_1
$$

$$
+ \int_{\frac{Q}{p}}^\infty\left(\frac{Q}{x}\right)f(t_1)\,dt_1. \tag{10}
$$

The stochastic failure and repair rate follow continuous distribution and we have considered exponential distributions as,

$$
F(t_1) = 1 - e^{-\phi t_1}, \qquad F(t_2) = 1 - e^{-\sigma t_2}.
$$

With exponential failure and repair time distributions, Equation (9) and Equation (10) becomes,

$$
ETC(Q, p, \phi) = S_m\left(1 + \frac{J}{\phi}\right) + \frac{C_{elc}e_{rs}}{\phi} + C_{elc}e_s + A\alpha\left(\frac{p - p_{min}}{p_{max} - p}\right)^\beta + R_m e^{\frac{\zeta\phi}{\varrho}} + \frac{e^{-\frac{Q\phi}{p}}\left(e^{\frac{Q\phi}{p}} - 1\right)}{\phi}
$$

$$
\left(c_p + C_{elc}e_1 + C_{elc}e_2 p + L\phi + Mp + p^2\omega + pr\phi\right) + \frac{(C_h + C_{elc}e_h)(p - x)e^{-\frac{Q\phi}{p}}}{x\phi^2}\left(p\left(e^{\frac{Q\phi}{p}} - 1\right)\right.
$$

$$
\left.- Q\phi\right) - \frac{x^2\phi C_s\left(e^{\frac{Q\left(-\frac{p\sigma}{x} + \sigma - \phi\right)}{p}} - 1\right)}{\sigma(p\sigma + x(\phi - \sigma))} \tag{11}
$$

and

$$E[T](Q,p,\phi) = \left(e^{-\frac{Q\phi}{p}-\frac{Q\sigma}{x}}\left(x^2\phi^2 e^{\frac{Q\sigma}{p}} - \left(p\sigma^2(p-x) + p\sigma x\phi + x^2\phi^2\right)e^{\frac{Q\phi}{p}+\frac{Q\sigma}{x}}\right.\right.$$
$$\left.\left.p\sigma + e^{\frac{Q\sigma}{x}}(p\sigma + x(\phi-\sigma))\right)\right) \bigg/ \sigma x\phi(x(\sigma-\phi)-p\sigma). \tag{12}$$

Now, the expected cost per unit time is given using renewal reward theorem [52], as

$$\mathrm{ETC}(Q,p,\phi) = t_1 \overset{\lim}{\to} \infty \frac{E\left[\text{total cost }[0,t_1]\right]}{t_1} = \frac{E[\mathrm{TC}](Q,p,\phi)}{E[T](Q,p,\phi)}. \tag{13}$$

3.4. Solution Methodology

In this study, production lot size Q, controllable production rate p, and manufacturing reliability ϕ are considered decision variables to optimize the expected total cost of system under stochastic production capacities and random energy consumption. The expected total cost of the system is a nonlinear constrained problem, whereas the goal of this paper is to optimize Equation (13) under the productivity and reliability constraints,

$$-Q \leq 0, \quad x < p, \quad p \leq p_{\max},$$
$$\phi_{\min} - \phi \leq 0, \quad \phi - \phi_{\max} \leq 0.$$

The mathematical problem is solved with Kuhn–Tucker method. Let $\lambda_1, \lambda_2, \lambda_3, \lambda_4, \lambda_5$ be the Lagrange multipliers corresponding to the above-mentioned constraints. Then, the Kuhn–Tucker (KT) necessary conditions for optimality are given as,

$$\frac{\partial ETC}{\partial Q} + \lambda_1 = 0, \frac{\partial ETC}{\partial p} - \lambda_2 + \lambda_3 = 0, \frac{\partial ETC}{\partial \phi} - \lambda_4 + \lambda_5 = 0. \tag{14}$$

$$\lambda_1 Q = 0, \tag{15}$$

$$\lambda_2(x-p) = 0, \quad \lambda_3(p_{\max}-p) = 0, \tag{16}$$

$$\lambda_4(\phi - \phi_{\max}) = 0, \quad \lambda_5(\phi_{\min} - \phi) = 0, \tag{17}$$

$$\lambda_1, \lambda_2, \lambda_3, \lambda_4, \lambda_5, \geq 0. \tag{18}$$

To ensure the convexity of objective function, the sufficient conditions from Kuhn–Tucker must be satisfied as all principle minors should be greater than 0. Since developed equation system ETC^* for the stochastic model is a complex function with higher powers, hence it is not possible to prove its optimality through analytic validity of sufficient conditions.

4. Numerical Experiment

The developed model was tested with numerical analysis and sensitivity analysis of the key input parameters. The modified values of input data were obtained from [26,28] and are given in Table 2.

Table 2. General input parameter values for numerical example.

$p_{max} = 2500$ (units/unit time)	$c_p = 1700$ (\$)	$M = 10$ (\$/per unit)
$r = 2$ \$/per unit	$e_h = 8.34$ (kWh/unit/unit time)	$L = 1500$ (\$)
$\gamma = 1$	$C_{elc} = 0.12$ (\$/kWh)	$S_m = 1200$ (\$/setup)
$e_s = 5835$ (kWh/setup)	$J = 0.0725$	$C_s = 5$ (\$/lost unit)
$h = 1.49$ (\$/unit/unit time)	$\omega = 0.0015$	$\phi_{max} = 1$
$\phi_{min} = 0$	$e_r = 1700$ (kWh/unit time)	$\sigma = 4$ (repairs/unit time)
$C_r = 500$ (\$/time)	$\zeta = 1.1$	$x = 300$ (units/unit time)
$\varrho = 0.5$	$R_m = 1500$ (\$)	$A = 200$ \$
$p_{min} = 300$ (units/unit time)	$\alpha = 1.1$	$\beta = 0.01$
$e_{rs} = 5835$ (kWh/setup)	$e_1 = 8500$ (kWh/unit time)	$e_2 = 50$ (kWh/unit)

4.1. Example

The optimal results for the integrated production–maintenance policy for an unreliable manufacturing system under energy control variables is achieved as, $ETC^* = 9253.75$ (\$/unit time) $Q^* = 1637.4$ (units) $p^* = 1095.72$ (units/unit time), $\phi^* = 0.24$.

The generated Hessian matrix for Equation (13) is given below. Moreover, the Hessian matrix at optimal values of decision variables is positive definite, given as,

$$
H = \begin{bmatrix}
\frac{\partial^2 ETC}{\partial Q^2} & \frac{\partial^2 ETC}{\partial Q \partial \phi} & \frac{\partial^2 ETC}{\partial Q \partial p} \\
\frac{\partial^2 ETC}{\partial \phi \partial Q} & \frac{\partial^2 ETC}{\partial \phi^2} & \frac{\partial^2 ETC}{\partial \phi \partial p} \\
\frac{\partial^2 ETC}{\partial p \partial Q} & \frac{\partial^2 ETC}{\partial p \partial \phi} & \frac{\partial^2 ETC}{\partial p^2}
\end{bmatrix}
= \begin{bmatrix}
0.000917294 & 0.536973 & 0.000349362 \\
0.536973 & 25447.1 & -0.995246 \\
0.000349362 & -0.995246 & 0.000889964
\end{bmatrix}
$$

and all eigenvalues (25447.1, 0.00124988, 0.000507122 of the matrix are positive.

The impact of required specific energy consumption for production on model parameters were tested for limits of -70% to $+100\%$. The graphical analysis illustrated in Figure 7 shows the impact of fixed and variable component of required specific energy for production (e_1, e_2) on optimal production quantity Q, production rate p, reliability parameter ϕ and expected total cost of system ETC. Figure 7 shows that the expected total cost of the system is extremely sensitive to variable component of specific energy requirement.

Moreover, as the value of energy components increases, there is a constant increase in all model parameters. In the case of constant component e_1 of specific energy consumption, the unit production rate increases with an increasing value of e_1, as this part is equally distributed over the total produced units. The production lot Q decreases with an increase of e_1 to reduce total cost of system. Similarly, the increases in reliability parameter suggest that, as the total cost of system increases with e_1, increased value of reliability parameter reduces the required technology development cost. On the other hand, the variable component of required specific energy e_2 shows a more sensitive relation to total cost of system and optimal production rate. Higher values of e_2 lead to a decrease in production rate to optimize the total cost of system. Similarly, the increase in e_2 leads to increase in value of reliability parameter to decrease the technology development cost for the manufacturing system.

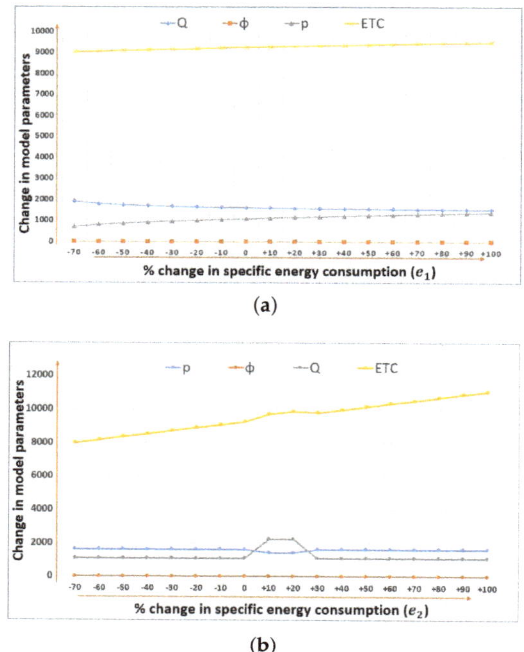

Figure 7. Change in Model parameters (ETC, Q, p, ϕ) of the system with respect to specific production energy parameters: (**a**) with varying fixed component of required energy for manufacturing (e_1); and (**b**) with varying variable component of required energy for manufacturing (e_2).

4.2. Sensitivity Analysis

A sensitivity analysis was performed for all key parameters and the results are compiled in Tables 3 and 4. Sensitivity of basic model parameters related to manufacturing reliability under energy consumption is given in Table 3.

- Higher setup costs S_m increase the total cost of manufacturing system but decrease the production rate of system. The proposed model also suggests increase in production quantity Q for higher setup costs to avoid multiple setup costs. On the other hand, the restoration cost J shows an increasing pattern for production quantity, production rate and expected total cost of the system. As the restoration cost depends on two parameters, setup cost and manufacturing reliability of system, the increasing value of restoration cost suggest higher production lot to avoid multiple setup requirement. In addition, it suggests higher production rates to shorten the production up-time to avoid added costs of longer production runs and possible system failure.

- For the decreasing shortage cost C_s, the model shows insensitivity to production lot size. However, in the case of an increase in shortage cost in the model, we see an increase in manufacturing reliability and decrease in the production lot size.

- Inventory holding cost h shows no impact on manufacturing reliability ϕ in this model. However, an increase in holding cost suggests an increase in production rate along with reduction of production quantity to reduce inventory accumulation. Higher inventory holding costs lead to higher system costs. The energy cost e_h for inventory holding shows the same pattern as inventory holding cost. With decreasing energy requirements for inventory carrying, the model suggests an increase in production lot size and production rate. Manufacturing reliability is insensitive to energy requirements for inventory holding.

- The technology development cost R_m has direct impact on the manufacturing reliability and productivity of system. The manufacturing reliability is defined as a system-design variable and it represents the number of failures a system faces during operational hours. As the increase in technology investment R_m shows a reduction in reliability parameter, it also reduces the production rate of system to adjust the total optimal cost of system. The production quantity increases with higher development costs, as it shows a more reliable manufacturing system.
- With the increase in corrective maintenance cost C_r, the model suggests an increase in production rate to adjust the optimal total cost of the system. Higher corrective maintenance cost leads to smaller production quantities, and the expected total cost of system also increases due to higher maintenance costs. The increasing value of σ suggest an increase in production quantity; as $\frac{1}{\sigma}$ represents the mean time to repair, as the mean time to repair decreases, the idle time for manufacturing system decreases. In addition, the model shows increase in production rate and the expected total cost of system; as the productivity of manufacturing system increases, the machine is in operational state for longer periods of time.
- As the required energy cost for system repair e_r increases, the model shows a reduction in production quantity but increase in production rate. The reduction in production lot size explains that, as total energy required for maintenance increases, the specific energy consumption per unit for repair increases. Therefore, the model suggest an increase in production rate to optimize the production policy under energy consumption. The total expected cost also increases with an increase in required electrical energy for repair activities.

Table 4 provide a comprehensive analysis of key parameters of model under energy consumption and stochastic production capacities. From the results, the following important insights are obtained:

- Unit production cost c_p comprises labor and specific energy consumption costs required to produce a production quantity; as the unit manufacturing cost decreases, we see an increase in production lot size of the model. In addition, with decrease in unit manufacturing cost, the expected total cost and production rate decreases. For both labor and production energy cost E_p, as the production rate decreases, the optimal reliability parameter also decreases, which is due to lower stress on the manufacturing system. Moreover, as production lot increases, reliability decreases to adjust technology development cost.
- The raw material cost M and raw material yield loss cost r both are highly sensitive to any increasing values of model parameters. The increase in material yield loss cost increases the production quantity and production rate simultaneously to adjust the total expected cost. In addition, it shows a reduction in the optimal reliability parameter of the model that is quite obvious; as for the manufacturing system, where raw material yield loss adds significant cost to manufacturing, higher reliability parameter is optimal for manufacturing system.
- The increase in tool/die cost ω shows a reduction in optimal production rate. As the production rate of manufacturing system increases, the added load/stress on manufacturing system increases the maintenance requirement of such machinery, and, to adjust the higher cost of production-technology development, the model suggests reduction in reliability value and optimal production rate.
- The manufacturing yield loss L due to manufacturing system failures also increases the expected total cost of the system. Moreover, the increase in manufacturing yield loss cost suggests a decrease in production quantity and justifies increasing the production rate to improve the productivity of the system. The increase in yield loss value suggests reducing the manufacturing reliability parameter to minimize the negative effect of manufacturing system failures.
- As the unit cost A to increase production rate by one unit decreases, the model shows an increase in production rate. With the increase in the production rate, the reliability parameter also increases. As the higher stress value increases failure frequency of manufacturing system, reliability parameter increases accordingly to adjust the technology development cost to keep

total cost at an optimal limit. The productivity constants show quite unique trends; as the value of productivity scale parameter α increases, the production rate decreases continuously with an increasing trend in production quantity and expected total cost of system. However, it is insensitive towards manufacturing reliability parameter. The productivity shape parameter β is insensitive towards expected total cost and reliability parameter of the system, however an increase in shape parameter increases the production quantity and decreases the production rate monotonically.

Table 3. Sensitivity analysis for parameters related to manufacturing reliability.

Parameter	% Change	Q	ϕ	p	% Change in ETC
S_m	-20	1611.36	0.24	1100.05	-0.55
	-10	1615.32	0.24	1099.61	-0.46
	$+10$	1654.56	0.24	1095.44	0.47
	$+20$	1673.92	0.24	1093.45	0.90
J	-20	1628.46	0.23	1094.99	-0.21
	-10	1631.75	0.23	1096.24	-0.11
	$+10$	1638.27	0.24	1098.75	0.11
	$+20$	1442.67	0.27	2253.6	3.54
C_s	-20	1635.00	0.23	1097.20	0.00
	-10	1635.01	0.23	1097.36	0.00
	$+10$	1435.99	0.27	2262.39	3.38
	$+20$	1436.01	0.27	2262.40	3.38
h	-20	1736.89	0.24	1134.34	-2.32
	-10	1683.17	0.24	1115.96	-1.14
	$+10$	1587.60	0.24	1077.01	1.21
	$+20$	1552.56	0.24	1060.60	2.17
e_s	-20	1635.01	0.23	1097.30	0.00
	-10	1635.02	0.23	1097.42	0.00
	$+10$	1435.98	0.27	2262.38	3.38
	$+20$	1435.99	0.27	2262.39	3.38
e_h	-20	1700.41	0.24	1122.09	-1.53
	-10	1666.54	0.24	1109.82	-0.76
	$+10$	1605.64	0.24	1085.11	0.74
	$+20$	1578.21	0.24	1072.63	1.46
σ	-20	1633.61	0.24	1095.13	-0.01
	-10	1634.41	0.24	1096.58	-0.01
	$+10$	1635.51	0.24	1098.08	0.00
	$+20$	1635.90	0.24	1099.45	0.00
C_r	-20	1635.50	0.24	1096.00	-0.02
	-10	1635.29	0.24	1096.66	-0.01
	$+10$	1634.76	0.24	1098.33	0.01
	$+20$	1634.50	0.24	1098.33	0.02
e_r	-20	1635.23	0.23	1096.83	-0.01
	-10	1635.13	0.24	1097.16	-0.00
	$+10$	1634.92	0.24	1097.83	0.00
	$+20$	1634.82	0.24	1098.16	0.01
R_m	-20	1550.14	0.26	1129.66	-1.55
	-10	1593.41	0.26	1113.23	-0.76
	$+10$	1675.26	0.26	1082.32	0.73
	$+20$	1710.49	0.26	1069.04	1.36
ζ	-20	1578.13	0.26	1131.53	-0.80
	-10	1623.86	0.24	1104.14	-0.16
	$+10$	1670.69	0.23	1076.42	0.50
	$+20$	1689.53	0.22	1065.35	0.77
ϱ	-20	1698.76	0.21	1059.96	0.89
	-10	1668.95	0.23	1077.44	0.48
	$+10$	17,278	0.05	300.06	-7.55
	$+20$	22,244	0.05	300.17	-7.76

Table 4. Sensitivity analysis for parameters related to production capacity.

Parameter	% Change	Q	ϕ	p	% Change in *ETC*
c_p	−20	1700.58	0.23	962.96	−1.32
	−25	1663.05	0.23	1034.12	−0.64
	+10	1436.44	0.27	2257.64	3.66
	+20	1594.67	0.24	1209.48	1.18
e_1	−20	1672.19	0.23	1018.60	−0.62
	−10	1653.42	0.23	1058.36	−0.31
	+10	1623.49	0.24	1131.16	0.30
	+20	1608.94	0.24	1166.59	0.58
e_2	−20	1637.45	0.23	1096.31	−3.89
	−10	1637.42	0.23	1096.02	−1.94
	+10	1439.66	0.24	2240.77	4.99
	+20	1439.76	0.27	2238.81	6.54
r	−20	1629.01	0.24	1105.86	−0.39
	−10	1631.99	0.24	1101.65	−019
	+10	1638.11	0.24	1093.39	0.19
	+20	1439.43	0.26	2246.64	3.74
M	−20	1635.11	0.24	1098.46	−8.03
	−10	1635.07	0.24	1097.98	−4.02
	+10	1634.98	0.24	1097.01	4.02
	+20	1436.76	0.27	2249.79	11.36
ω	−20	1583.03	0.25	1248.08	−1.41
	−10	1608.94	0.24	1166.81	−0.68
	+10	1637.65	0.24	1091.10	0.07
	+20	1688.38	0.23	984.278	1.25
L	−20	1640.96	0.24	1079.23	−0.24
	−10	1638.25	0.24	1087.43	−0.13
	+10	1631.98	0.23	1107.38	0.13
	+20	1629.09	0.23	1117.08	0.26
A	−20	1627.73	0.24	1079.23	−0.24
	−10	1632.59	0.24	1087.43	−0.13
	+10	1638.06	0.23	1107.38	0.13
	+20	1641.09	0.23	1117.08	0.26
α	−20	1629.25	0.24	1098.87	−0.12
	−10	1632.59	0.24	1098.07	−0.05
	+10	1638.06	0.24	1096.77	0.06
	+20	1641.09	0.24	1096.05	0.13
β	−20	1634.95	0.24	1097.71	0.00
	−10	1634.99	0.24	1097.60	0.00
	+10	1635.06	0.24	1097.39	0.00
	+20	1635.01	0.24	1097.28	0.00

4.2.1. Comparative Study

This section provides a comprehensive analysis for the influence of decision variables on the expected total cost of the system. As shown in Figure 8a, as production lot size increases, the total expected cost of the system decreases to an optimal production quantity where total cost is minimum. The graph shows that *ETC* is not monotonically decreasing in Q, as there exists an optimal production lot size Q, after which the total expected costs starts to increase with an increase in production quantity.

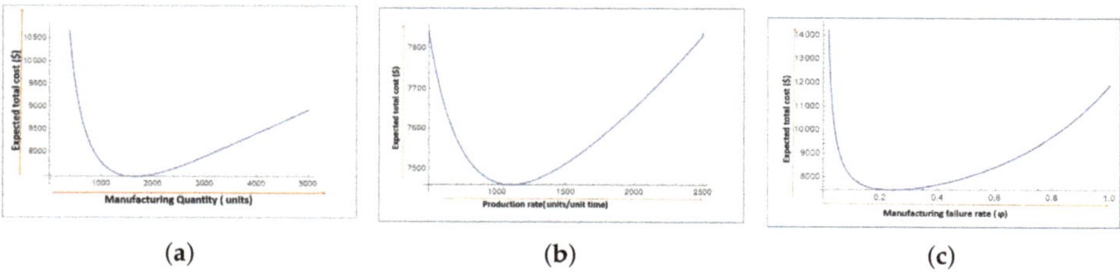

(a)   (b)   (c)

Figure 8. Change in expected total cost $ETC(Q, p, \phi)$ of the system with respect to decision variables: (**a**) varying production quantity (Q); (**b**) varying production rate (p); and (**c**) varying reliability parameter (ϕ).

Figure 8b depicts the effect of controllable production rate p on the expected total cost of manufacturing system ETC. With the increase in production rate, the expected total cost decreases to a certain limit as per unit production cost $C(p, \phi)$ is a function of production rate, after which it starts to increase again. The higher production rate also results in lower specific energy consumption per unit of production as more items are produced in less production up-time. However, increase in production rate increases other cost components of per unit production cost, i.e., tool/die cost, thus we see an increase in total cost after the optimal limit for model, which is 1095.72 (units/unit time). The expected total cost ETC monotonically increases with manufacturing reliability parameter ϕ, as shown in Figure 8c. As the failure rate of manufacturing system increases, the total cost increases due to productivity loss, and extra costs of repair, energy consumption, and resulted idle time in manufacturing process.

As shown in Figure 9, the expected unit production cost $C(p, \phi)$ monotonically increases with manufacturing unreliability. As the failure rate of manufacturing system increases, the expected unit production cost increases due to productivity loss, extra costs of repair and idle time. The production rate shows a rather different relation with expected unit production cost. As the production rate increases, the expected unit production cost decreases to an optimal point, after which it starts to increase.

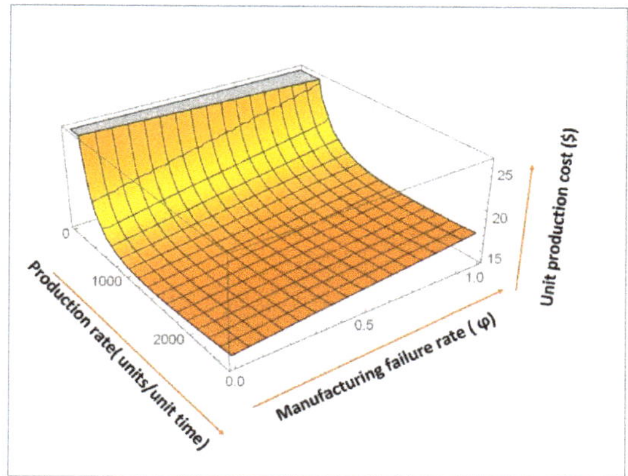

Figure 9. Change in expected unit manufacturing cost $c(p, \phi)$ with varying manufacturing reliability (ϕ) and production rate (p).

As the repair rate increases, time for each repair t_2 decreases and eventually energy consumption per repair decreases, which leads to a decrease in repair cost for manufacturing system, as shown in Figure 10.

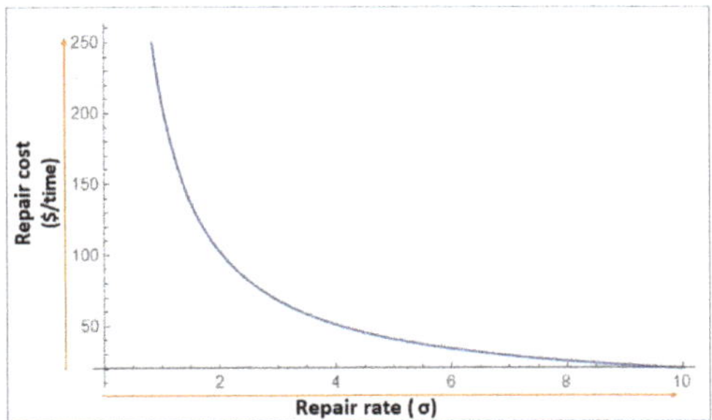

Figure 10. Change in energy consumption per machine repair $SEC(\sigma)$ with varying repair rate (σ).

Figure 11a shows the relationship of the expected total cost ETC with manufacturing unreliability. As the failure rate of manufacturing system increases, the expected unit production cost also increases. With an increase in unit production rate, the expected total cost of system decreases because the unit production cost is the sum of manufacturing cost and energy consumption for production and, with an increase in production rate, the energy cost decreases [44]. Figure 11b shows the relationship of the expected total cost ETC with manufacturing quantity and production rate. The manufacturing quantity shows a convex relationship with expected total cost of system. Figure 11c shows the relationship of the expected total cost ETC with manufacturing quantity and manufacturing reliability parameter. The manufacturing quantity shows a convex relationship with expected total cost of system. ETC increases monotonically with increase in failure rate of manufacturing system.

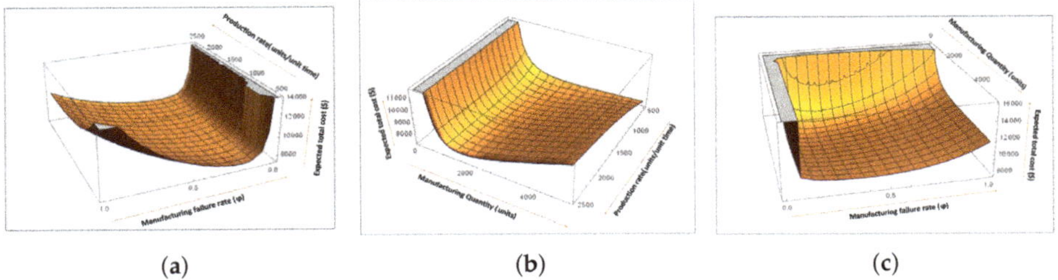

(a)   (b)   (c)

Figure 11. Change in expected total cost $ETC(Q, p, \phi)$ of the system with respect to decision parameters: (**a**) varying manufacturing reliability parameter and production rate $(\phi,\ p)$; (**b**) varying manufacturing quantity and production rate $(Q,\ p)$; and (**c**) varying manufacturing quantity and manufacturing reliability parameter $(Q,\ \phi)$.

4.2.2. Managerial Insights

- The study provides a better understanding of production–maintenance policy for stochastic production capacities under random failure rates, random repair rates and energy consumption requirements.
- By employing precise cost data of related cost factors associated with unreliable manufacturing systems, production and maintenance managers can set an optimum level of manufacturing reliability that minimizes the overall maintenance and energy consumption costs.
- Random breakdowns of manufacturing system represent an inevitable phenomenon, which is considered in many maintenance planning models for energy consumption, but the effect of random repair rate is usually ignored. This study also gives insights into the simultaneous improvement of reliability and manufacturing productivity when managers faces the challenge of uncertain production capacities due to longer repair times.

- The increase in tool/die cost shows a reduction in optimal production rate. As the production rate of manufacturing system increases, the added load and stress increases the maintenance requirement of such machinery, and to adjust the higher cost of production-technology development, the model suggests reductions in reliability value and processing speeds.
- For controllable production rates, the added expenditure of maintenance and technology development investment are also considered; therefore, managers can also look into possible technology development needed to improve productivity of manufacturing systems.
- This paper presents the relevant relationships between variable energy consumption rates and constituents related to stochastic production capacities and maintenance management plans. Improved production and maintenance planning will allow managers to increase production rates and reduce energy consumption per unit of production quantity. Furthermore, through additional production and technology development policies, machine breakdowns can be reduced.

5. Conclusions

This study investigated an integrated production–maintenance–inventory system with some realistic assumptions. For an unreliable manufacturing system, the manufacturing capacity is considered stochastic as it depends on the random failure rate, repair rate and production rate of manufacturing system, which follow exponential distributions. Hence, the energy consumption for each production cycle is also considered random for unreliable manufacturing systems. The developed stochastic model is optimized for production lot size, production rate and reliability parameter. The practical implications of model are shown through sensitivity analysis and comparative study of model parameters.

- The study suggested that it is not practical to consider constant production rates in an unreliable manufacturing system; as the system faces random failure rate and random repair rate, the productivity of system is highly affected due to manufacturing unreliability. In such a scenario, controllable production rates help in increasing the production capacity of a manufacturing system.
- As the productivity of a manufacturing system depends on its reliability and operational condition, the unit production cost of a manufacturing system is considered as a function of controllable production rate, which varies within design limits, manufacturing reliability, labor costs and variable energy costs.
- The manufacturing system productivity and reliability can be further enhanced through investment in advance technology, better and improved resources and skilled labor. Therefore, a production-technology development cost is introduced as a function of manufacturing reliability. For improved efficiency of manufacturing process, a productivity investment is also introduced in manufacturing costs to get optimal level of variable processing speeds for an unreliable manufacturing system.
- The cost of electrical energy consumption for simultaneous production, inventory and maintenance planning is not considered in the existing literature, therefore this study introduces the energy cost components for operational and idle states of production setups, restoration, production, inventory holding and repair cost.
- A limitation of the study is considering deterministic demand, where random demand of products is a more realistic phenomena. In addition, only electrical energy consumption is considered in this study, whereas production firms usually use more than one type of energy sources in production. The model can be further extended with the case of imperfect production process, inspection policies and economic factors such as inflation and time value of money. Moreover, integrated production–maintenance policies under uncertain production environments are worthy of investigation, as discussed in [53]. The study can be further extended to the concept [54] of

imperfect production process for an unreliable manufacturing system, as production of defectives is inevitable.

Author Contributions: Conceptualization, I.A. and S.K.; Methodology, I.A.; Software, I.A.; Validation, B.S.; Formal analysis, I.A; Investigation, B.S. and I.A; Resources, B.S., I.A., and S.K.; Data curation, B.S., I.A and S.K.; Writing—original draft preparation, I.A; Writing—review and editing, I.A; Visualization, I.A. and S.K.; Supervision, B.S.

Funding: This work was supported by reserach fund of Hanyang University (HY-2018-F), under (Project Number: 201800000001370).

Conflicts of Interest: The authors declare no conflicts of interest.

Abbreviations

The following abbreviations are used in this paper.

EMQ	Economic manufacturing quantity
O2O	Online to Offline
AR	Abort/Resume
KT	Kuhn–Tucker method
SEC	Specific energy consumption
EEPP	Energy efficient production planning

References

1. Sarkar, B.; Guchhait, R.; Sarkar, M.; Pareek, S.; Kim, N. Impact of safety factors and setup time reduction in a two-echelon supply chain management. *Robot. Comput.-Integr. Manuf.* **2019**, *55*, 250–258. [CrossRef]
2. Ciavotta, M.; Meloni, C.; Pranzo, M. Minimising general setup costs in a two-stage production system. *Int. J. Prod. Res.* **2013**, *51*, 2268–2280. [CrossRef]
3. Kumar, V.; Sarkar, B.; Sharma, A.N.; Mittal, M. New product launching with pricing, free replacement, rework, and warranty policies via genetic algorithmic approach. *Int. J. Comput. Int. Syst.* **2019**, *12*, 519–529. [CrossRef]
4. Sarkar, B.; Guchhait, R.; Sarkar, M.; Cárdenas-Barrón, L.E. How does an industry manage the optimum cash flow within a smart production system with the carbon footprint and carbon emission under logistics framework? *Int. J. Prod. Econ.* **2019**, *213*, 243–257. [CrossRef]
5. Shibin, K.; Gunasekaran, A.; Papadopoulos, T.; Childe, S.J.; Dubey, R.; Singh, T. Energy sustainability in operations: An optimization study. *Int. J. Adv. Manuf. Technol.* **2016**, *86*, 2873–2884. [CrossRef]
6. Bazan, E.; Jaber, M.Y.; Zanoni, S. Supply chain models with greenhouse gases emissions, energy usage and different coordination decisions. *Appl. Math. Model.* **2015**, *39*, 5131–5151. [CrossRef]
7. Giret, A.; Trentesaux, D.; Prabhu, V. Sustainability in manufacturing operations scheduling: A state of the art review. *J. Manuf. Syst.* **2015**, *37*, 126–140. [CrossRef]
8. Liu, Y.; Dong, H.; Lohse, N.; Petrovic, S.; Gindy, N. An investigation into minimising total energy consumption and total weighted tardiness in job shops. *J. Clean. Prod.* **2014**, *65*, 87–96. [CrossRef]
9. Biel, K.; Glock, C.H. Systematic literature review of decision support models for energy-efficient production planning. *Comput. Ind. Eng.* **2016**, *101*, 243–259. [CrossRef]
10. Liu, Y.; Dong, H.; Lohse, N.; Petrovic, S. A multi-objective genetic algorithm for optimisation of energy consumption and shop floor production performance. *Int. J. Prod. Econ.* **2016**, *179*, 259–272. [CrossRef]
11. Rapine, C.; Penz, B.; Gicquel, C.; Akbalik, A. Capacity acquisition for the single-item lot sizing problem under energy constraints. *Omega* **2018**, *81*, 112–122. [CrossRef]
12. Mousavi, S.; Thiede, S.; Li, W.; Kara, S.; Herrmann, C. An integrated approach for improving energy efficiency of manufacturing process chains. *Int. J. Sustain. Eng.* **2016**, *9*, 11–24. [CrossRef]
13. González-Romera, E.; Ruiz-Cortés, M.; Milanés-Montero, M.I.; Barrero-González, F.; Romero-Cadaval, E.; Lopes, R.A.; Martins, J. Advantages of minimizing energy exchange instead of energy cost in prosumer microgrids. *Energies* **2019**, *12*, 719. [CrossRef]
14. Mori, M.; Fujishima, M.; Inamasu, Y.; Oda, Y. A study on energy efficiency improvement for machine tools. *CIRP Ann.* **2011**, *60*, 145–148. [CrossRef]

15. Shrouf, F.; Ordieres-Meré, J.; García-Sánchez, A.; Ortega-Mier, M. Optimizing the production scheduling of a single machine to minimize total energy consumption costs. *J. Clean. Prod.* **2014**, *67*, 197–207. [CrossRef]
16. Tang, D.; Dai, M.; Salido, M.A.; Giret, A. Energy-efficient dynamic scheduling for a flexible flow shop using an improved particle swarm optimization. *Comput. Ind.* **2016**, *81*, 82–95. [CrossRef]
17. Dai, M.; Tang, D.; Giret, A.; Salido, M.A.; Li, W.D. Energy-efficient scheduling for a flexible flow shop using an improved genetic-simulated annealing algorithm. *Robot. Comput.-Integr. Manuf.* **2013**, *29*, 418–429. [CrossRef]
18. Li, X.; Xing, K.; Wu, Y.; Wang, X.; Luo, J. Total energy consumption optimization via genetic algorithm in flexible manufacturing systems. *Comput. Ind. Eng.* **2017**, *104*, 188–200. [CrossRef]
19. Calicioglu, O.; Flammini, A.; Bracco, S.; Bellù, L.; Sims, R. The Future Challenges of Food and Agriculture: An integrated analysis of trends and solutions. *Sustainability* **2019**, *11*, 222. [CrossRef]
20. Pramangioulis, D.; Atsonios, K.; Nikolopoulos, N.; Rakopoulos, D.; Grammelis, P.; Kakaras, E. A Methodology for Determination and Definition of Key Performance Indicators for Smart Grids Development in Island Energy Systems. *Energies* **2019**, *12*, 242. [CrossRef]
21. Brahman, F.; Honarmand, M.; Jadid, S. Optimal electrical and thermal energy management of a residential energy hub, integrating demand response and energy storage system. *Energy Build.* **2015**, *90*, 65–75. [CrossRef]
22. Shojafar, M.; Cordeschi, N.; Baccarelli, E. Energy-efficient adaptive resource management for real-time vehicular cloud services. *IEEE Tran. Cloud Comput.* **2016**, *7*, 196–209. [CrossRef]
23. Sarkar, B.; Tayyab, M.; Choi, S.B. Product Channeling in an O2O supply chain management as power transmission in electric power distribution systems. *Mathematics* **2019**, *7*, 4. [CrossRef]
24. Bazan, E.; Jaber, M.Y.; Zanoni, S. Carbon emissions and energy effects on a two-level manufacturer-retailer closed-loop supply chain model with remanufacturing subject to different coordination mechanisms. *Int. J. Prod. Econ.* **2017**, *183*, 394–408. [CrossRef]
25. Bazan, E.; Jaber, M.Y.; El Saadany, A.M. Carbon emissions and energy effects on manufacturing-remanufacturing inventory models. *Comput. Ind. Eng.* **2015**, *88*, 307–316. [CrossRef]
26. Sarkar, M.; Sarkar, B.; Iqbal, M. Effect of energy and failure rate in a multi-item smart production system. *Energies* **2018**, *11*, 2958. [CrossRef]
27. Devoldere, T.; Dewulf, W.; Deprez, W.; Willems, B.; Duflou, J.R. Improvement potential for energy consumption in discrete part production machines. In *Advances in Life Cycle Engineering for Sustainable Manufacturing Businesses*; Springer: London, UK, 2007; pp. 311–316.
28. Sarkar, B.; Sana, S.S.; Chaudhuri, K. Optimal reliability, production lotsize and safety stock: An economic manufacturing quantity model. *Int. J. Manag. Sci. Eng. Manag.* **2010**, *5*, 192–202. [CrossRef]
29. Sarkar, B.; Sana, S.S.; Chaudhuri, K. Optimal reliability, production lot size and safety stock in an imperfect production system. *Int. J. Math. Oper. Res.* **2010**, *2*, 467–490. [CrossRef]
30. Sarkar, B. An inventory model with reliability in an imperfect production process. *App. Math. Comput.* **2012**, *218*, 4881–4891. [CrossRef]
31. Bhuniya, S.; Sarkar, B.; Pareek, S. Multi-product production system with the reduced failure rate and the optimum energy consumption under variable demand. *Mathematics* **2019**, *7*, 465. [CrossRef]
32. Yao, X.; Zhou, J.; Zhang, J.; Boër, C.R. From intelligent manufacturing to smart manufacturing for Industry 4.0 driven by next generation artificial intelligence and further On. In Proceedings of the 2017 5th International Conference on Enterprise Systems (ES), Beijing, China, 22–24 September 2017; IEEE: Piscataway, NJ, USA, 2017; pp. 311–318.
33. Sarkar, B.; Mandal, P.; Sarkar, S. An EMQ model with price and time dependent demand under the effect of reliability and inflation. *App. Math. Comput.* **2014**, *231*, 414–421. [CrossRef]
34. Glock, C.H. Batch sizing with controllable production rates. *Int. J. Prod. Res.* **2010**, *48*, 5925–5942. [CrossRef]
35. Shepherd, R.W. *Theory of Cost and Production Functions*; Princeton University Press: Princeton, NJ, USA, 2015.
36. Rishel, R. Control of systems with jump Markov disturbances. *IEEE Trans. Autom. Control* **1975**, *20*, 241–244. [CrossRef]
37. Olsder, G.; Suri, R. Time-optimal control of parts-routing in a manufacturing system with failure-prone machines. In Proceedings of the 1980 19th IEEE Conference on Decision and Control including the Symposium on Adaptive Processes, Albuquerque, NM, USA, 10–12 December 1980; IEEE: Piscataway, NJ, USA, 1980; pp. 722–727.

38. Glock, C.H. Batch sizing with controllable production rates in a multi-stage production system. *Int. J. Prod. Res.* **2011**, *49*, 6017–6039. [CrossRef]
39. Khouja, M.; Mehrez, A. Economic production lot size model with variable production rate and imperfect quality. *J. Oper. Res. Soc.* **1994**, *45*, 1405–1417. [CrossRef]
40. Sarkar, M.; Hur, S.; Sarkar, B. Effects of variable production rate and time-dependent holding cost for complementary products in supply chain model. *Math. Probl. Eng.* **2017**, *2017*, 2825103. [CrossRef]
41. Alfares, H.K.; Ghaithan, A.M. EOQ and EPQ Production-Inventory Models with Variable Holding Cost: State-of-the-Art Review. *Arab. J. Sci. Eng.* **2019**, *44*, 1737–1755. [CrossRef]
42. Dey, B.K.; Sarkar, B.; Pareek, S. A two-echelon supply chain management with setup time and cost reduction, quality improvement and variable production rate. *Mathematics* **2019**, *7*, 328. [CrossRef]
43. Khalifehzadeh, S.; Fakhrzad, M. A Modified Firefly Algorithm for optimizing a multi stage supply chain network with stochastic demand and fuzzy production capacity. *Comput. Ind. Eng.* **2019**, *133*, 42–56. [CrossRef]
44. Marchi, B.; Zanoni, S.; Jaber, M.Y. Economic production quantity model with learning in production, quality, reliability and energy efficiency. *Comput. Ind. Eng.* **2019**, *129*, 502–511. [CrossRef]
45. Marchi, B.; Zanoni, S.; Zavanella, L.; Jaber, M. Supply chain models with greenhouse gases emissions, energy usage, imperfect process under different coordination decisions. *Int. J. Prod. Econ.* **2019**, *211*, 145–153. [CrossRef]
46. Adane, T.; Nicolescu, M. Towards a Generic Framework for the Performance Evaluation of Manufacturing Strategy: An Innovative Approach. *J. Manuf. Mater. Process.* **2018**, *2*, 23. [CrossRef]
47. Demichela, M.; Baldissone, G.; Darabnia, B. Using field data for energy efficiency based on maintenance and operational optimisation. A step towards PHM in process plants. *Processes* **2018**, *6*, 25. [CrossRef]
48. Rackow, T.; Kohl, J.; Canzaniello, A.; Schuderer, P.; Franke, J. Energy Flexible Production: Saving Electricity Expenditures by Adjusting the Production Plan. *Procedia CIRP* **2015**, *26*, 235–240. [CrossRef]
49. Lee, S.; Prabhu, V.V. Energy-aware feedback control for production scheduling and capacity control. *Int. J. Prod. Res.* **2015**, *53*, 7158–7170. [CrossRef]
50. Gutowski, T.; Dahmus, J.; Thiriez, A. Electrical energy requirements for manufacturing processes. In Proceedings of the 13th CIRP International Conference on Life Cycle Engineering, Leuven, Belgium, 31 May–2 June 2006; Volume 31, pp. 623–638.
51. Prevention, I.P. *Control Reference Document on Best Available Techniques for Large Combustion Plants*; European Commission: Brussels, Belgium, 2006.
52. Ross, S.M. *Introduction to Probability Models*; Academic Press: Cambridge, MA, USA, 2014.
53. Dellino, G.; Kleijnen, J.P.; Meloni, C. Robust optimization in simulation: Taguchi and response surface methodology. *Int. J. Prod. Econ.* **2010**, *125*, 52–59. [CrossRef]
54. Sarkar, B.; Majumder, A.; Sarkar, M.; Kim, N.; Ullah, M. Effects of variable production rate on quality of products in a single-vendor multi-buyer supply chain management. *Int. J. Adv. Manuf. Technol.* **2018**, *99*, 567–581. [CrossRef]

Article

Effect of Electrical Energy on the Manufacturing Setup Cost Reduction, Transportation Discounts, and Process Quality Improvement in a Two-Echelon Supply Chain Management under a Service-Level Constraint

Irfanullah Khan [1], Jihed Jemai [2], Han Lim [1] and Biswajit Sarkar [3,*]

[1] Department of Industrial & Management Engineering, Hanyang University, Ansan, Gyeonggi-do 15588, Korea; irfanullah13@outlook.com (I.K.); nanhan2@hanyang.ac.kr (H.L.)
[2] Department of Industrial Engineering, Hanyang University, Seoul 04763, Korea; jihed.jemei@gmail.com
[3] Department of Industrial Engineering, 50 Yonsei-ro, Sinchon-dong, Seodaemun-gu, Seoul 03722, Korea
* Correspondence: bsarkar@yonsei.ac.kr or bsbiswajitsarkar@gmail.com; Tel.: +82-107-498-1981

Received: 22 July 2019; Accepted: 21 September 2019; Published: 29 September 2019

Abstract: The need for efficient electrical energy consumption has greatly expanded in the process industries. In this paper, efforts are made to recognize the electrical energy consumption in a two-echelon supply chain model with a stochastic lead-time demand and imperfect production, while considering the distribution free approach. The initial investments are made for quality improvement and setup cost reduction, which ultimately reduce electrical energy consumption. The inspection costs are considered in order to ensure the good qualities of the product. Centralized and decentralized strategies are used to analyze the proposed supply chain model. The main objective of this study is to reduce the overall cost through efficient electrical energy consumption in supply chain management by optimizing the lot size, the number of shipments, the setup cost, and the failure rate. A quantity-based transportation discount policy is applied to reduce the expected annual costs, and a service-level constraint is incorporated for the buyer to avoid a stockout situation. The impact of the decision variables on the expected total costs is analyzed, and sensitivity analysis is carried out. The results show a significant reduction in overall cost, with quality improvement and setup cost reduction ultimately reducing electrical energy consumption.

Keywords: electrical energy consumption; supply chain management; imperfect production; distribution free approach; service level constraint; transportation discount

---

1. Introduction

The supply chain is a complex process consisting of several parameters that need to be controlled. Among them, the few parameters that can be controlled by making some additional investments include the setup cost, lead time, and quality of the products, as shown by Sarkar and Sarkar [1]. Certain parts of the manufacturing units need to be improved with state-of-the-art technology to reduce the energy consumption and enhance the quality of the products. These parts are replaced with some additional investment for setup cost and process quality improvement. Managers seek to optimize the supply chain costs; therefore, they analyze various aspects of the supply chain management systems. In the process industries, the cost of energy consumption is usually a small portion of the overall production cost. Therefore, less attention is given to energy consumption. Among the various types of energies used in the supply chain operation, electrical energy is the most consumed. Thus, efforts are made in this model to recognize the cost associated with electrical energy consumption.

In the supply chain management literature, most of the supply chain models have considered a fixed setup cost, constant lead time, and the perfect quality of the products, as indicated by Sarkar et al. [2]. However, the manufacturing setup cost can be reduced, the process quality can be improved, and the lead time can be controlled by making smart investments in various aspects of the system. The production system moves from an in-control state to an out-of-control state due to the following reasons: persistent usage of the machinery, inadequate process control, human negligence, and mishandling during the shipment. Therefore, imperfect-quality products are manufactured with the perfect-quality products in a long-run production system. These imperfect products add some additional cost to the vendor in the shape of replacement costs for defective items, and during this long process, an enormous amount of energy is consumed. In the imperfect production process, a huge amount of energy is consumed during the inspection process, defective unit replacement, ordering time, and setup time. In such situations, with the enormous amount of energy consumption, the risk of product shortage may occur, which may lead to damage to the brand image.

A number of quality management practices have been reported in the production management literature to deal with the uncertainties in the production systems. In this course of action, Salameh and Jaber [3] extended the basic economic production quantity (EPQ) model to an imperfect production system by incorporating a full inspection of the entire lot, but there is not a single assumption in their model related to energy consumption during the production. The basic inventory model is extended to the case of an imperfect production process in several studies, and effective polices are provided for the overall system cost reduction. However, these studies did not consider efficient energy consumption in the production system [4–6]. Several models in this direction considered the defective rate as a constant, such as those in the research of Salameh and Jaber [3], Khan et al. [7], and Chung [8]. However, in many studies, the random defective rate is considered (see Tayyab and Sarkar [9] and Cárdenas-Barrón et al. [10]). Moreover, the emergence of philosophies such as "just-in-time" in the recent past has enabled managers to focus on the reduction of lead-time components. These components include the order preparation time, transit time, lead time of the supplier, time of delivery, and the setup time. In fact, the lead time is reduced by incurring an additional amount as the crashing cost on lead-time components. These costs include administrative costs, agile transportation, and the vendor's process speed-up cost for the reduction in waiting time. There are several advantages of the reduction of the lead time in the form of a high service level, quick response time, and—due to the lead-time reduction—less energy being consumed. This can provide an advantage to overcome the uncertainties of product demand in a competitive environment (see Ben-Daya and Hariga Sarkar [5]).

Based on the above perspective, the objective of this study is to design an optimization model with the objective to reduce the total costs of the system while considering the imperfect production and the efficient electrical energy consumption. A two-echelon supply chain model with two strategies is developed in this study. The first model is developed based on a centralized framework, while the second model provides a decentralized approach. In the decentralized policy, the interactive decisions between the retailer and the manufacturer are based on the Stackelberg approach. The imperfect production process is considered with a variable lead time, and the buyer's lead time is reduced by adding a crashing cost. Additional initial investments are made to improve the process quality and reduce the setup cost. These investments are used to update the technologies, which helps produce good quality products with less energy consumption; in turn, this will reduce the overall supply-chain costs. The initial investment made for quality improvement retains the production system in the in-control state, which saves energy consumption due to this state having a smaller number of production system failures than the existing one. The service-level constraint is used to meet the demand from the existing stock. The transportation discounts are allowed on a quantity basis, and the optimal lot size is determined. The distribution-free approach (Scarf [11] and Gallego and Moon [12]) is used to solve the model.

In this model, costs related to shortages are not considered for the buyers because of the high penalty costs associated with shortages. Due to the stockout situations, the buyer can face the worst situation in the form of goodwill or brand image loss, and if the shortage occurs during the production process, an enormous amount of energy is wasted to backorder. In the inventory management and operation management literature, it is suggested that the cost associated with shortages and service-level constraints have an equal impact on the overall cost of the model simultaneously. Due to the use of service-level constraint, the consumption of electrical energy reduces. Therefore, the service-level constraint is used as the replacement of shortages in the system. In the literature, the shortage cost is replaced with the condition on service level, and it is assumed that the buyer has set a certain service level in terms of the fill rate that satisfies the corresponding proportion of demand from the available stock [13–16]. Therefore, the service-level constraint sets a limit on the proportion of demand that was not met from the available stock, which should not exceed a specified value. Consequently, the buyer service level measure reflects that a certain fraction of the demand, which is satisfied regularly, is expressed by λ.

The basic inventory model by Harris [17] assumes that arriving goods are of perfect quality, although the supply process of goods rarely conforms to the specification. This results in the deviation in quantity or time, which leads to delay and might also negatively affect the brand image. To meet the inferior quality manifestations, many researchers relaxed the perfect supply assumption made in the basic inventory model. Considering the inventory model, Salameh and Jaber [3] analyzed the lot size with imperfect quality items and entire lot inspection. During the screening period in their model, defective items are collected (in batches), and those batches are sold in secondary markets that did not consider the cost associated with the energy consumption during the production. Cárdenas-Barrón [18] provided the corrected solution for the Salameh and Jaber [3] model. Ciavotta et al. [19] minimized the general setup cost within a two-stage production system. They organized the production in batches by considering identical jobs. A robust methodology is developed in the model of Dellino et al. [20] for an uncertain environment. They explained the classical economic ordered quantity with the robust optimization technique. However, no one has considered the amount and associated cost of energy consumption in their models.

To determine the imperfect items in the production model, Goyal and Cárdenas-Barrón [21] discussed a simple approach with optimal results compared to the Salameh and Jaber model [3]. Huang [22] proposed an extension to the Salameh and Jaber [3] model by integrating the inventory model and considering a single-supplier single-buyer integrated-inventory model with an imperfect production system along with inspection activity during the just-in-time environment. Their extension was amazing in the inventory field. However, with the passage of time and increasing competition, certain other parameters need to be recognized, as the energy cost is not much compared to other cost. Still, it possesses some value, which they did not recognize in their model. There are some other additional parameters in this model, such as the introduction of transportation discounts, lead-time reduction, and service-level constraints, which improve the process of the efficient utilization of energy. Ben-Daya and Hariga [5] developed an integrated inventory model to obtain the optimal lot size, the reorder point with imperfect items, and the number of shipments. They considered that the lead demand follows a normal distribution and depends on the lot size and the transportation delay. A vendor–buyer integrated inventory model was developed by Dey and Giri [6] considering the random lead-time demand, the product inspection, and the quality improvement with the shortages. Kim et al. [23] extended the integrated inventory model of Ben-Daya and Hariga [5], and proposed an improved way of calculating imperfect items with backorder items. Still, no one recognizes the cost associated with the electrical energy consumption in their modeling. The efficient electrical energy utilization can only be done once the importance of the energy is recognized.

The inventory model with a finite replenishment rate was developed by Sarkar [24] for the retailer considering the permissible delay-in-payments and stock dependent demand. The imperfect multi-stage lean production system is considered with a rework and random defective rate in the research of Tayyab and Sarkar [9]; however, in their model, rework is performed on the defective items. They did not consider the initial investment for quality improvement in an imperfect production system; by rework, additional costs are incurred on the products, and sometimes, rework is not possible for quality improvement. No author has thought about the cost associated with electrical energy consumption during defective product production. The increase in energy-related costs, including its demand, legislation, and its consumer's environmental awareness in recent times has influenced the industries to consider carefully the energy consumption and the associated cost. Therefore, the recent studies related to the production model have considered the energy-related costs. Tang et al. [25] investigated the production planning problem in a stochastic environment, and considered the energy consumption costs for the steel production process. In their model, the energy consumption cost was the nonlinear function of production quantity. Gahm et al. [26] developed the energy-efficient scheduling approaches to improve the energy efficiency. Furthermore, they introduced three energy dimensions: energy supply, energy demand, and energetic coverage. Keller and Reinhart [27] considered an approach that integrated energy supply information in the enterprise resource planning system, and they had the idea of energy flexibility, energy efficiency, and industry internal and external energy supply.

Du et al. [28] and Todde et al. [29] developed the models considering energy consumption and energy analysis. In the Tomić and Schneider [30] model, the energy recovery methods were explained from the waste by applying a closed-loop energy recovery approach. Haraldsson and Johansson [31] studied the various types of the energy efficiency in a production system; however, their model did not consider the imperfect production system. Although energy is consumed throughout the production system, these energy consumption-related costs are usually low, which is why due importance is generally not given to them compared with the other production-related costs. Specially, the main gap in the existing literature is in the supply chain model, where the energy-related costs are not considered. These costs have an enormous impact on overall cost as well as environmental impact. This study initiates the use of energy consumption cost in a two-echelon supply with an imperfect production process.

Managers dealing with production decisions under uncertain conditions must know the lead-time distribution of the demand. However, it's very challenging and time-consuming to find the accurate distribution pattern. Therefore, it's crucial for industries and researchers to find a way of calculating the total system cost without having the data on lead-time demand distribution. To deal with this problem, Scarf [11] first introduced the distribution-free newsboy issue with a known standard deviation and mean of the lead-time demand. Gallego and Moon [12] modified the ordering rule of [11]. Scarf [11] initiated the ordering rule when the distribution of lead-time demand is not present. Gallego and Moon [12] simplified the approach developed by scarf [11]. This approach is used when distribution does not follow any pattern, but the variance and the mean of the demand are available. It was used for finding the optimal ordering quantity to maximize the profit in case of the worst possible distribution of the demand.

The distribution-free approach is used in many directions, specifically in the inventory field. To find the optimal ordering quantity and the reordering point, a distribution-free continuous review inventory model was developed by Moon and Choi [14]. The optimal ordering quantity and lead time were considered by Ouyang and Wu [13]. They introduced the lead-time crashing cost. Ouyang and Chang [32] utilized the logarithmic function of Porteus [33] for quality improvement in the imperfect production process by applying the distribution-free procedure. To reduce the setup cost and to improve the quality, Ouyang et al. [4] utilized the distribution-free approach in a lot size reorder point model with the imperfect production process. The continuous review inventory model by Ma and Qiu [16] considered the distribution-free approach with an improvement in the setup cost while considering a service-level constraint. A transportation discount was introduced by Shin et al. [34] in a review inventory model with controllable lead time and a service-level constraint. The joint replenishment model was provided by Braglia et al. [35] considering stochastic lead-time demand with backorder and lost sales, but they did not consider the efficient energy consumption and cost associated with energy.

The electrical energy consumption and the related costs in the imperfect production system have not been studied in the available supply chain management literature. The cost associated with electrical energy consumption is modeled in this paper. Nevertheless, these costs are a smaller portion of the production cost, but the energy is consumed in every supply chain operation, and there is a need to reduce the associated cost. The wastage of energy is not only a financial loss; it also damages the environment. The proposed study develops a two-echelon supply chain model with an imperfect production system and identifies the electrical energy consumption costs under a stochastic demand. To reach the global optimal solution, an improved algorithm is developed in this paper. The initial investments are made for quality improvement and setup cost reduction. The inspection process is performed at the buyer end to reduce the electrical energy consumption. The service-level constraint is considered, and the demand is met from the available stock. Transportation discounts are offered on a certain quantity to attract customers. The min–max distribution-free approach is considered in this model with a known mean and standard deviation regarding the lead-time demand. Generally, the information on lead-time demand distribution is difficult to collect. If the distribution is known, then the expected shortage amount can be determined by utilizing the identified distribution. If the lead-time distribution of the demand is not available, then the statistical procedures are used to calculate it. The distribution-free approach provides prominent results; therefore, this approach is utilized to develop the proposed model. Table 1 presents the contribution of this paper.

Table 1. Contributions of authors.

Author(s)	Supply Chain Management	Electrical Energy Consumption	Process Quality Improvement in Imperfect Production	Setup Cost Reduction	Distribution-Free Approach	Transportation Discount
Porteus [33]			✓	✓		
Gallego and Moon [12]					✓	
Moon and Choi [14]					✓	
Ouyang and Wu [13]					✓	
Ouyang et al. [4]			✓	✓	✓	
Ma and Qiu [16]				✓	✓	
Tsao and Lu [36]						✓
Ouyang et al. [37]	✓		✓			
Jha and Shanker [38]	✓					✓
Dey and Giri [6]	✓		✓			
Priyan and Uthayakumar [39]	✓			✓		✓
Tayyab and Sarkar [9]			✓			
Ben-Daya and Hariga [5]	✓					
Tang et al. [25]		✓				
Shin et al. [34]					✓	✓
Kim et al. [23]	✓		✓	✓	✓	
Gutgutia and Jha [40]	✓				✓	
This paper	✓	✓	✓	✓	✓	✓

2. Problem Definition, Notation, and Assumptions

This section consists of the problem definition, notations, and the assumptions of the model.

2.1. Problem Definition

A two-echelon supply chain model is considered with efficient energy consumption under an imperfect production process. In this model, some initial capital investments are made for the production process quality improvement and setup cost reduction. The investment, made in process quality improvement, is restricting the production process to move to the out-of-control state from the in-control state. When the production process moves to the out-of-control state, a huge amount of energy is wasted on defective products. Therefore, the initial investment is made for process quality improvement, and the setup cost reduction reduces the total cost of the supply chain with the efficient electrical energy consumption. A quantity-based transportation discount is offered to motivate buyers to increase their ordering quantity. Lead-time crashing cost is used for the buyer to reduce the lead time, which ultimately reduces the electrical energy consumption. The number of shipments and the smart lot size are optimized. The service-level constraint is used to avoid shortages, and the fraction of demand is met through the available inventory. Finally, the aim of this model is to minimize the overall supply chain costs with the recognition and minimum electrical energy consumption. The production lot, reorder point, and the number of shipments is optimized. The flow of the supply chain with costs associated to the vendors and buyers are shown in Figure 1.

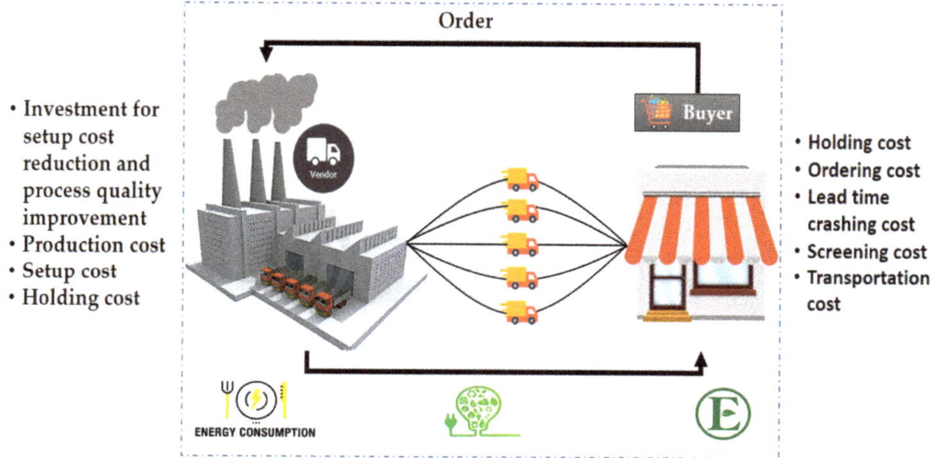

Figure 1. Supply chain flow diagram under efficient energy consumption.

2.2. Notation

This section describes the related notation of this model (Table 2).

Table 2. Notation for parameter and decision variables.

	Decision variable
Q	order quantity (units)
ϕ	probability of production process that may go to out-of-control state
S	setup cost for vendor per setup ($/setup)
n	number of shipments per lot (positive integer)
L	length of the lead time (weeks)
r	reorder point

Table 2. *Cont.*

	Parameters
P	production rate (units/year)
ρ	reciprocal of P (years/unit)
x	screening rate (units/year)
D	average demand per year (units/year)
σ	standard deviation of demand (units/week)
$E(x)$	expected value of x
X	lead-time demand that has a probability distribution function F
x^{+}	max $[x, o]$
$E(X - r)^{+}$	expected shortage per replenishment cycle
μ	mean of the lead-time demand
A	ordering cost of the buyer ($/order)
A'	cost of the energy consumed per order of the buyer ($/order)
C	crashing cost per order ($/order)
C'	cost for amount of energy consumed during lead-time crashing ($/order)
S_0	initial setup cost for vendor per lot ($/setup)
S'	cost of energy consumed during setup ($/setup)
L_0	initial lead time length in weeks (weeks)
h_{b1}	holding cost for buyer's defective products per unit per year ($/unit/year) $h_{b1} < h_{b2}$
h'_{b1}	cost of energy consumed during buyer's defective products holding ($/unit/year)
h_{b2}	holding cost for buyer's non-defective products per unit per year ($/unit/year),
h'_{b2}	cost of energy consumed during buyer's non-defective products holding ($/unit/year)
h_v	holding cost for vendor's per unit per year ($/unit/year)
h'_v	cost of energy consumed during the vendor holding ($/unit/year)
s	screening cost per unit ($/unit)
s'	energy cost consumed during screening ($/unit)
W	defective unit replacement cost ($/defective unit)
W'	energy cost consumed during defective unit replacement ($/defective unit)
α	fractional annual cost of capital investment ($/year)
λ	the fraction of customers' regular satisfied demand
ϕ_0	the initial probability of the production process shifting from control state to out-of-control state
b	coefficient to control quality improvement capital investment cost
B	coefficient of capital investment for controlling setup cost reduction
u_j	minimum duration for j^{th} lead-time component (days), $j = 1, 2, \ldots, n$
v_j	normal duration for j^{th} lead-time component (days), $j = 1, 2, \ldots, n$
m_i	crashing cost per day for ith component of lead time ($/day), $i = 1, 2, \ldots, n$
m'_i	cost of energy consumed during crashing for ith component ($/day), $i = 1, 2, \ldots, n$
m_j	crashing cost per day for j^{th} component of lead time with m_j as ($/day), $j = 1, 2, \ldots, n$
m'_j	cost of energy consumed during crashing for j^{th} component ($/day), $j = 1, 2, \ldots, n$
γ_i	transportation cost, $i = 1, 2, \ldots, n$
γ'_i	energy consumed during transportation, $i = 1, 2, \ldots, n$

2.3. Assumptions

The following assumptions are considered during the formulation of the proposed supply chain management model:

1. In this supply chain model, the cost of electrical energy consumed during the entire operations (e.g., during setup, ordering, inspection, and inventory holding time) is considered for a single type of product.
2. The electrical energy consumption is taken care of thoroughly with continuous review of the buyer's inventory; when the inventory level falls to the reorder point, the replenishments are made.
3. When the buyer receives the lot, it is assumed that the complete lot is inspected with the screening rate x under efficient electrical energy consumption. Furthermore, the assumption is made that the inspection process is non-destructive and error-free in nature.
4. To omit the shortages from the model, a service-level constraint is utilized with a specific fill rate.
5. The initial capital investments are made for process quality improvement and setup cost reduction by which advanced equipment can be used that consumes the minimum amount of electrical energy.
6. To avoid the extra cost of electrical energy incurred due to the shortages, the supplier production rate and the screening rate are higher than the demand rate. The assumption indicates that $Q(1+\phi)(1-\phi) \geq \frac{DQ(1+\phi)}{x}$, which can be simplified as $\phi \leq 1 - D/x$ (refer to Salameh and Jaber [3]). Figure 2 shows the flow of inventory in relation to inspection time.

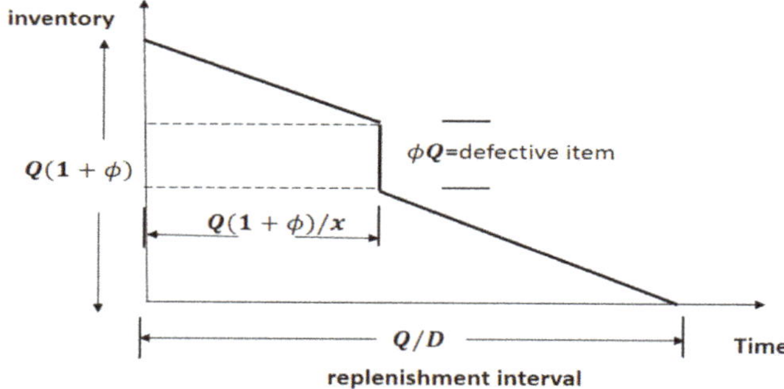

Figure 2. Inventory flow diagram.

3. Mathematical Model

The proposed mathematical model is formulated in this section for a two-echelon supply chain under energy consumption consideration.

3.1. Vendor's Model

This section formulates the vendor's inventory model as in the Ben-Daya and Hariga [5] model while considering the energy consumption cost in the supply chain. As in most of the supply chain models, the energy consumption cost is not considered during the process, because it is usually a small portion of supply chain cost. Nevertheless, the energy cost has a certain impact on the overall cost. The energy is consumed during various operations such as heating, cooling, lightening, etc. In this model, the electrical energy consumption is considered throughout the entire supply chain. The vendor expected annual cost comprises production setup cost with electrical energy consumption during setup time, the inventory-holding cost with electrical energy consumption, the defective unit replacement cost with energy consumption, the initial investment for setup cost reduction, and process quality improvement. This model is developed for the vendor's production situation where the buyer places the order, and the vendor starts producing the ordered quantity until the production run has been completed. As under the single-setup multi-delivery policy, the vendor produces the items in the quantity of nQ, and the buyer receives these products in lots of size Q. Therefore, the production cycle length of the vendor is $\frac{nQ}{D}$.

To reduce the supply chain cost, in this model, two initial capital investments are used for product quality improvement and setup cost reduction. The concept of Porteus [33] is considered to improve the production process quality. Initially, the production process is considered as the in-control state; with the passage of time, the production process may move to the out-of-control state, where the defective units are produced until the entire lot is produced, which consumes a significant amount of energy. One good strategy that can resist the system from moving to the out-of-control state from the in-control state is an additional capital investment, which is able to reduce the energy consumptions due to the failure of the production system. For instance, to reduce the out-of-control system probability from 0.00002 to 0.000018, an investment of \$200 is made; then, another \$200 is furtherly required to reduce it down to 0.000016, and so on. Ultimately, making an initial investment a good choice for the reduction of imperfect production.

Production setup cost with electrical energy consumptions cost

The vendor's setup cost per production setup is S; as under the single-setup multi-delivery policy, the vendor's production cycle length is $\frac{nQ}{D}$. A certain amount of energy is consumed during the production setup time, the energy cost incurred during setup is S' per setup. Therefore, the vendor's production setup cost is:

$$= \frac{The\ vendor's\ setup\ cost\ per\ production\ setup + the\ energy\ consumuption\ during\ setup}{The\ expected\ production\ cycle\ length} = \frac{D(S + S')}{nQ}$$

Warranty cost with electrical energy consumptions cost

In this model, the $Q\phi$ defective units are replaced with the cost of W per product and W' energy consumed during the defective unit replacement, and the vendor's production cycle length is $\frac{Q}{D}$; therefore, the annual cost of defective unit replacement is:

$$= (W + W')D\phi$$

Process quality improvement with electrical energy consumptions cost

The capital investment $I(\phi)$ is made (as Porteus [33]) to improve the process quality with a reduction in the out-of-control probability as follows:

$$I(\phi) = ab\ ln\ ln\left(\frac{\phi_0}{\phi}\right),\ 0 < \phi \le \phi_0.$$

If the investment function, $I(\phi) = 0$, it implies that no capital investment is made for the quality improvement. However, if some investments are made, then the value of ϕ_0 will decrease, which indicates that the product quality improvement has been completed with the capital investment. In the actual production system, the ϕ value is considered low; this is the reason that the low value of ϕ is used in this model. The advantage of applying the logarithmic function is its convexity for the established range of investment function.

Setup cost reduction with electrical energy consumptions cost

In most of the basic inventory models, a fixed setup cost is used. The setup cost of the model can be reduced by using initial investments. Initially, the investment can be important, but it will be reduced in each phase of the model. For this purpose, the logarithmic investment function is used (refer to Porteus [33]):

$$I(S) = \alpha B \ln \ln \left(\frac{S_0}{S}\right), \quad 0 < S \le S_0$$

Vendor's holding cost with electrical energy consumptions cost

The average inventory of the vendor can be calculated (from Figure 3) as the difference of the vendor's accumulated inventory and the buyer's accumulated inventory.

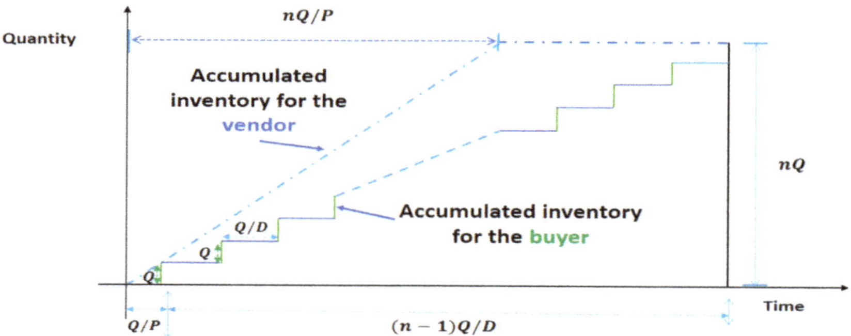

Figure 3. Inventory pattern for the vendor.

That is:

$$\text{Vendor holding cos t per year} = \frac{[\textit{doted area}-\textit{bold area}]h_v D}{nQ}$$

$$= \frac{\left[nQ\left(\frac{Q}{P}+\frac{(n-1)Q}{D}\right)-nQ\left(\frac{nQ}{P}\right)\left(\frac{1}{2}\right)-\frac{Q}{D}\{Q+2Q+3Q+\ldots+(n-1)Q\}\right]h_v D}{nQ},$$

inserting $\frac{1}{P} = \rho$ in the above equation:

$$= \left[\left\{nQ\left(Q\rho+(n-1)\left(\frac{Q}{D}\right)\right)-\frac{n^2Q^2\rho}{2}\right\}-\left\{\left(\frac{Q^2}{D}\right)(1+2+3+\ldots+(n-1))\right\}\right]\left(\frac{h_v D}{nQ}\right),$$

simplifying by the sum of sequence formula:

$$= \left[\left\{nQ\left(Q\rho+(n-1)\left(\frac{Q}{D}\right)\right)-\frac{n^2Q^2\rho}{2}\right\}-\left\{\left(\frac{Q^2}{D}\right)\left(\frac{n(n-1)}{2}\right)\right\}\right]\left(\frac{h_v D}{nQ}\right)$$

Further simplification of the above equation:

$$= h_v\left(\frac{Q}{2}\right)[n(1-D\rho)-1+2D\rho]$$

Therefore, the vendor's holding cost per unit time is:

$$= h_v\left(\frac{Q}{2}\right)[n(1-D\rho)-1+2D\rho].$$

As a certain amount of electrical energy is consumed during the holding time, the cost associated with the energy during the holding time is h'_v. Therefore, the vendor's holding cost became:

$$= (h_v + h'_v)\left(\frac{Q}{2}\right)[n(1 - D\rho) - 1 + 2D\rho].$$

The vendor's expected annual cost with the recognition of electrical energy consumption cost is shown in Equation (1), and comprises the process quality improvement, setup cost reduction, setup cost, holding cost of the vendor, and defective cost:

$$\begin{aligned}
EAC'_v(Q, S, \phi, n) = \quad & ab\,ln\,ln\left(\frac{\phi_0}{\phi}\right) + \alpha B\,ln\,ln\left(\frac{S_0}{S}\right) + \frac{D(S+S')}{nQ} \\
& + \frac{(h_v + h'_v)Q}{2}[n(1 - D\rho) - 1 + 2D\rho] + (W + W')D\phi \\
& \text{for } 0 < \phi \leq \phi_0 \text{ as well as } 0 < S \leq S_0
\end{aligned} \tag{1}$$

3.2. Buyer's Model

In this section, a buyer's inventory model is developed based on Ben-Daya and Hariga [5] study. The two-echelon supply chain model with imperfect production and efficient energy consumption is considered. In this continuous review (Q, r) inventory model, the vendor produces nQ quantities to save the buyer's holding cost. The quantity produced in a production cycle is delivered in n shipments to the buyer. The buyer places an order of quantity Q of non-defective items from the vendor as soon as the inventory level falls to reorder point r, where $r = DL + k\sigma\sqrt{L}$, and DL is the expected demand during lead time; k is the safety factor that fulfills the probability that the buyer's lead-time demand exceeds the reorder point r; and $k\sigma\sqrt{L}$ is the safety stock (see Hadley and Whitin, [41] and Ben-Daya and Hariga [5]). The vendor replenishes the order quantity after the lead time. However, due to the imperfect production system in the batch, some items are defective; thus, the buyer cannot sell all the replenished items. Hence, the buyer inspects the replenished products to separate the defective products from the non-defective products. This paper considered a 100% inspection of the buyer with screening rate x to locate the imperfect items in the lot, as mentioned in assumption 4. It is also assumed that the screening process is error-free and non-destructive. These defective products are identified, separated out, and returned to the vendor in the next lot.

The defective items can be reduced to some level, but cannot be eliminated entirely. These defective items create unexpected shortages in the system. To avoid the losses due to defective items in this model, the vendor sends an additional amount of inventory to the buyer, unlike in the previous studies of Huang [22] and Dey and Giri [6]. In the literature, the demand from the buyer is met by reducing the replenishment interval to $\frac{(1-\phi)Q}{D}$ (see [3,18,22]). However, the replenishment interval is not reduced in this model; to avoid the shortages due to imperfect products, the buyer orders additional products from the vendor. Consider the situation in which the buyer orders a quantity $Q + \phi Q + \phi^2 Q + \phi^3 Q + \ldots = \frac{Q}{(1-\phi)}$. The order quantity is approximated to $Q + \phi Q$ because ϕ has a negligible value. Consequently, the vendor gives $Q + \phi Q$ products to the buyer. This additional amount will fulfill the buyer's demand and simultaneously compensate for the losses due to the imperfect production process. The flow of material is shown in Figure 4. The buyer has two types of holding cost for defective and non-defective items in this model.

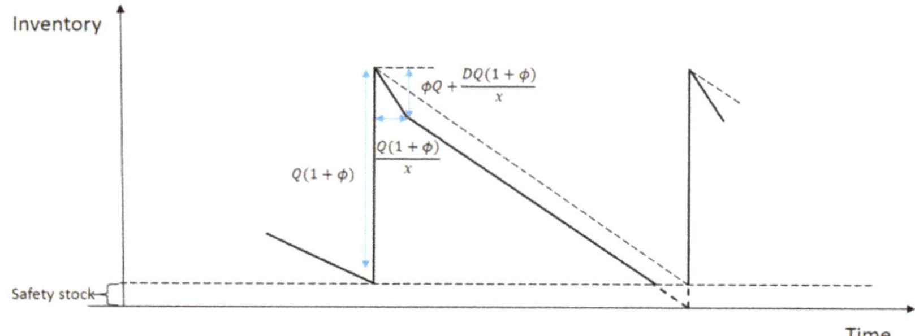

Figure 4. Inventory of the buyer with inspection.

Buyer holding cost for defective products with energy consumption

The amount of defective products that the buyer receives is $Q + \phi^2 Q$; the defective rate is ϕ. The defective items' quantity is approximated to ϕQ due to the very small value of ϕ. The number of defective items found is ϕQ during the inspection time $\dfrac{(1+\phi)Q}{x}$. It shows that the amount of average inventory of defective items per cycle is $\dfrac{\phi(1+\phi)Q^2}{2x}$. The average inventory of the identified defective products per year is $\phi Q - \dfrac{\phi(1+\phi)DQ}{2x}$. The cost incurred on holding the average inventory of the defective product per year of the buyer is h_{b1}, and the electrical energy consumed during the holding period is h'_{b1}. Therefore, the holding cost of the buyer is:

$$\left(h_{b1} + h'_{b1}\right)\left[\phi Q - \frac{\phi(1+\phi)DQ}{2x}\right]$$

Buyer holding cost for non-defective products with energy consumption

The amount of non-defective products that the buyer receives is $Q + \phi Q - \left(\phi Q + \phi^2 Q\right)$. The expected inventory of the non-defective product during the screening time is $\dfrac{Q}{2} + r - DL + \dfrac{\phi(1+\phi)DQ}{2x}$. The holding cost incurred on the non-defective products per year is h_{b2}, and the cost of the energy consumed during holding these products is h'_{b2}. Thus, the buyer holding cost for non-defective products is:

$$\left(h_{b2} + h'_{b2}\right)\left[\frac{Q}{2} + (r - \mu) + \frac{\phi(1+\phi)DQ}{2x}\right]$$

Buyer screening cost with energy consumption

As $Q(1 + \phi)$ products are screened with a per-unit screening cost of s and the cost of electrical energy consumed during screening being s'. Therefore, the annual screening cost is:

$$D(s + s')(1 + \phi).$$

Buyer ordering cost with energy consumption

The buyer ordering cost per order is A, and the energy consumed during the ordering time is A'. The buyer cycle length is $\dfrac{Q}{D}$. Thus, the ordering cost of the buyer is:

$$\frac{(A + A')D}{Q}.$$

Lead-time crashing cost with energy consumption cost

The lead time L consists of m mutually independent components, and these components include the order preparation time, order transit to the supplier, transit time from the supplier, setup time, and preparation time for availability. The jth component has a minimum duration of u_j, normal duration of v_j, and crashing cost per unit time of m_i, and the cost of energy consumed during the crashing time is m_i'. Furthermore, for convenience, it is assumed that $m_1 \leq m_2 \leq m_3, \ldots, \leq m_n$. It is clear that the lead-time reduction should start from the first component (the reason is the minimum crashing cost), and then the second component, and so on (Liao and Shyu [42], Ben-Daya and Hariga [5], and Kim et al. [23]). Let $L_0 = \sum_{j=1}^{n} v_j$ and L_i be the length of the lead-time components 1, 2, 3, $\ldots$, z that crashed in their minimum duration; then, L_i can be written as $L_i = \sum_{j=1}^{n} v_j - \sum_{j=1}^{i} (v_j - u_j)$, where $i = 1, 2, 3, \ldots$, n. The lead-time crashing cost per cycle with electrical energy consumption is $C(L)'$, as shown in Equation (2) (see Ouyang and Wu [13]) for a given $L \in [L_i, L_{i-1}]$:

$$C(L)' = (m_i + m_i')(L_{i-1} - L) + \sum_{j=1}^{i-1}(m_j + m_j')(v_j - u_j).$$

(2)

The buyer has to carry the additional costs incurred by the vendor on lead-time reduction in each stage of the replenishment cycle, and the lead-time crashing cost is also included in the buyer's cost.

Transportation discounts with electrical energy consumption cost

The concept of a transportation discount is incorporated in this paper. The cost of transportation is not constant, and it depends directly on the demand function. In the literature, the transportation cost is considered as fixed, i.e., independent of ordering quantity or shipment in most of the cases. In this study, the transportation cost is dependent on the demand as (see Priyan and Uthayakumar [39], Shin et al. [34]):

$$TC(Q) = \gamma_0 D, \ Q \in [M_0, M_1]$$
$$\gamma_1 D, \ Q \in [M_1, M_2]$$
$$\gamma_2 + D, Q \in [M_2, M_3]$$
$$\ldots$$
$$\gamma_N D, \ Q \in [M_N, \infty].$$

where $\gamma_0 > \gamma_1 > \gamma_2 \ldots > \gamma_N$, γ_i is the transportation cost per unit. M_i describes the range of $Q (M_0 = 0)$, i.e., Q should be in some specific range; else, $TC(Q)$ cannot be defined. Since the energy is consumed during transportation, the cost associated with the energy consumption during the transportation is γ_i'. Therefore, the transportation cost after considering the energy consumption cost will become as shown in Equation (3):

$$TC(Q)' = (\gamma_i + \gamma_i')D, \ Q \in [M_0, M_1].$$

(3)

The expected annual cost of the buyer (EAC_b), which consists of the ordering cost, lead-time crashing cost, holding cost for defective products, holding cost for non-defective products, screening cost, and the transportation discounts with electrical energy consumption during the process, is shown in Equation (4):

$$EAC_b'(Q, r, L) = \frac{D(A+A')}{Q} + \frac{DC(L)'}{Q} + \left(h_{b1} + h_{b1}'\right)\left[\phi Q - \frac{\phi(1+\phi)DQ}{2x}\right] + (h_{b2}$$
$$+ h_{b2}')\left[\frac{Q}{2} + (r - \mu) + \frac{\phi(1+\phi)DQ}{2x}\right] + D(s + s')(1 + \phi) + D\left(\gamma_i + \gamma_i'\right).$$

(4)

The service measure reflects that certain fractions of the demand, which is satisfied regularly from the available inventory, is expressed by λ.

The buyer service level in terms of the fill rate (λ) is described as shown in Equations (5) and (6):

$$\lambda = \frac{Expected\ demand\ satisfied\ per\ replenishment\ cycle}{Expected\ demand\ per\ replenishment\ cycle}, \tag{5}$$

$$\lambda = 1 - \frac{E[X-r]^+}{Q}, \tag{6}$$

which can be simplified as shown in Equation (7):

$$(1-\lambda)Q = E[X-r]^+. \tag{7}$$

To obtain the least favourable distribution F, the following lemma is used as shown in Equation (8) (Gallego and Moon [12]):

$$E[X-r]^+ \leq \frac{\sqrt{\sigma^2 L + k^2\sigma^2 L} - k\sigma\sqrt{L}}{2} \text{ for any } F \in F. \tag{8}$$

As $r = \mu + k\sigma\sqrt{L}$ or $r - \mu = k\sigma\sqrt{L}$ (see Hadley and Whitin, [41] and Ben-Daya and Hariga [5]), suppose that $y = r - \mu = k\sigma\sqrt{L}$, and it became as shown in Equation (9):

$$(1-\lambda)Q \leq \frac{\sqrt{\sigma^2 L + y^2} - y}{2}. \tag{9}$$

After solving Equation (9), it will become as shown in Equation (10):

$$y = \frac{\sigma^2 L}{4(1-\lambda)Q} - (1-\lambda)Q, \tag{10}$$

After inserting the value of y in Equation (10), it will become as shown in Equation (11):

$$r - \mu = \frac{\sigma^2 L}{4(1-\lambda)Q} - (1-\lambda)Q. \tag{11}$$

Inserting the value of $r - \mu$ in Equation (4), the annual expected cost of the buyer with energy consumption cost is shown in Equation (12):

$$\begin{aligned} EAC_b'(Q,r,L) = {} & \frac{D(A+A')}{Q} + \frac{DC(L)'}{Q} + \left(h_{b1} + h_{b1}'\right)\left[\phi Q - \frac{\phi(1+\phi)DQ}{2x}\right] + (h_{b2} \\ & + h_{b2}')\left[\frac{Q}{2} + \left(\frac{\sigma^2 L}{4(1-\lambda)Q} - (1-\lambda)Q\right) + \frac{\phi(1+\phi)DQ}{2x}\right] + D(s \\ & + s')(1+\phi) + D(\gamma_i + \gamma_i'). \end{aligned} \tag{12}$$

3.3. Supply Chain Cost

3.3.1. Centralized Analysis

The expected annual cost concerning the supply chain model with electrical energy consumption is sum of the vendor costs (as in Equation (1)) and the buyer's cost (as in Equation (4)). Accordingly, the EAC_s can be represented as shown in Equation (13):

$$
\begin{aligned}
EAC'_s(Q,S,\phi,n,r,L) \;=\;& \alpha b \, ln \, ln\left(\tfrac{\phi_0}{\phi}\right) + \alpha B \, ln \, ln\left(\tfrac{S_0}{S}\right) + \tfrac{D(S+S')}{nQ} + \tfrac{D(A+A')}{Q} + \tfrac{DC(L)'}{Q} \\
&+\left(h_{b1}+h'_{b1}\right)\left[\phi Q - \tfrac{\phi(1+\phi)DQ}{2x}\right] + (h_{b2} \\
&+h'_{b2})\left[\tfrac{Q}{2} + \left(\tfrac{\sigma^2 L}{4(1-\lambda)Q} - (1-\lambda)Q\right) + \tfrac{\phi(1+\phi)DQ}{2x}\right] + D(s \\
&+s')(1+\phi) + \tfrac{(h_v+h'_v)Q}{2}\left[n(1-D\rho)-1+2D\rho\right] + (W+W')D\phi \\
&+D\left(\gamma_i+\gamma'_i\right) \\
&\text{for } 0 < \phi \le \phi_0 \text{ as well as } 0 < S \le S_0.
\end{aligned}
\tag{13}
$$

In the above Equation (13), the cost components include: the investment made for quality improvement, the initial investments for the setup cost reduction, the setup cost with the energy consideration of the vendor, the ordering cost with the energy cost of the buyer, the holding cost of the defective items with the energy consideration of the buyer, the buyer non-defective item-holding cost, the screening cost with energy consideration, the holding cost of the vendor, the replacement cost with energy consideration, and the transportation discount cost with energy cost, respectively.

The necessary conditions to get the minimum cost of $EAC'_s(Q,n,S,\phi,r,L)$ with energy consideration are as follows in Equations (14)–(17):

$$
\begin{aligned}
\frac{\partial EAC'_s}{\partial Q} = \tfrac{1}{4}\Bigg[&-\left(\frac{4D\left((A+A')+C(L)'n+(S+S')\right)}{nQ^2}\right) + 2(h_v+h'_v)(n-1-D(n-2)\rho)+ \\
&4(h_{b1}+h'_{b1})\phi - \frac{2D(h_{b1}+h'_{b1})\phi(1+\phi)}{x} + (h_{b2}+h'_{b2})\left(-2+4\lambda+\frac{L\sigma^2}{Q^2(-1+\lambda)}+\frac{2D\phi(1+\phi)}{x}\right)\Bigg],
\end{aligned}
\tag{14}
$$

$$
\begin{aligned}
\frac{\partial EAC'_s}{\partial \phi} =\;& (h_{b1}+h'_{b1})Q + D((s+s')+(W+W')) - \frac{b\alpha}{\phi} - \\
&\frac{D\left((h_{b1}+h'_{b1})-(h_{b2}+h'_{b2})\right)Q(1+2\phi)}{2x},
\end{aligned}
\tag{15}
$$

$$
\frac{\partial EAC'_s}{\partial S} = \frac{D}{nQ} - \frac{B\alpha}{(S+S')},
\tag{16}
$$

$$
\frac{\partial EAC'_s}{\partial L} = \frac{D}{Q} + \frac{j\sigma^2}{4Q(1-\lambda)}.
\tag{17}
$$

The optimal values for the decision variables are calculated by using the necessary conditions as follows in Equations (18)–(21):

$$
Q'^* = \frac{\sqrt{-2D(n(A+A')+C(L)n+(S+S'))x(\lambda-1)+\tfrac{1}{2}h_{b2}Lnx\sigma^2)}}{\left(\sqrt{n(\lambda-1)\left(h_v x(1-n+D(n-2)\rho)+(h_{b1}+h'_{b1})\phi(D-2x+D\phi)-(h_{b2}+h'_{b2})(x(2\lambda-1)+D\phi(1+\phi))\right)}\right)},
\tag{18}
$$

$$
\phi'^* = \frac{1}{4A_1 DQ}\left(A_2 Dx - \sqrt{(A_3 + 16bDA_4)}\right),
\tag{19}
$$

where:

$$A_1 = \left((h_{b1} + h'_{b1}) - (h_{b2} + h'_{b2})\right)$$

$$A_2 = \left((h_{b2} + h'_{b2}) - (h_{b1} + h'_{b1})\right)Q + 2(h_{b1}Q + D((s + s') + (W + W')))$$

$$A_3 = \left(D\left((h_{b1} + h'_{b1}) - (h_{b2} + h'_{b2})\right)Q - 2\left((h_{b1} + h'_{b1})Q + D((s + s') + (W + W'))\right)x\right)^2$$

$$A_4 = \left((h_{b2} + h'_{b2}) - (h_{b1} + h'_{b1})\right)Qx\alpha)$$

$$S^* = \frac{BnQ\alpha}{D}, \tag{20}$$

$$y^* = \frac{\sigma^2 L}{4(1 - \lambda)Q} - (1 - \lambda)Q. \tag{21}$$

The optimal values of n are obtained by the expression shown in Equation (22):

$$EAC_s(n-1) \geq EAC_s(n) \leq EAC_s(n+1). \tag{22}$$

Sufficient conditions are satisfied by taking the second derivative of the decision variable with respect to the expected annual cost.

From the sufficing conditions, the following values are obtained as shown in Equations (23)–(25):

$$\frac{\partial^2 EAC'_s}{\partial Q^2} = \frac{2(A + A')D}{Q^3} + \frac{2C(L)'D}{Q^3} + \frac{2D(S + S')}{nQ^3} + \frac{(h_{b2} + h'_{b2})L\sigma^2}{2Q^3(1 - \lambda)}, \tag{23}$$

$$\frac{\partial^2 EAC_s}{\partial \phi^2} = -\frac{D(h_{b1} + h'_{b1})Q}{x} + \frac{D(h_{b2} + h'_{b2})Q}{x} + \frac{b\alpha}{\phi^2}, \tag{24}$$

$$\frac{\partial^2 EAC_s}{\partial S^2} = \frac{B\alpha}{(S + S')^2}, \tag{25}$$

To find the global minimum EAC'_s, the lemma is shown in Appendix A.

This Algorithm 1 is developed to determine the numerical solution of the proposed model, and the process flow of the algorithm is shown in Figure 5. Below is the detailed explanation of the algorithm to obtaining the optimal solution of the problem.

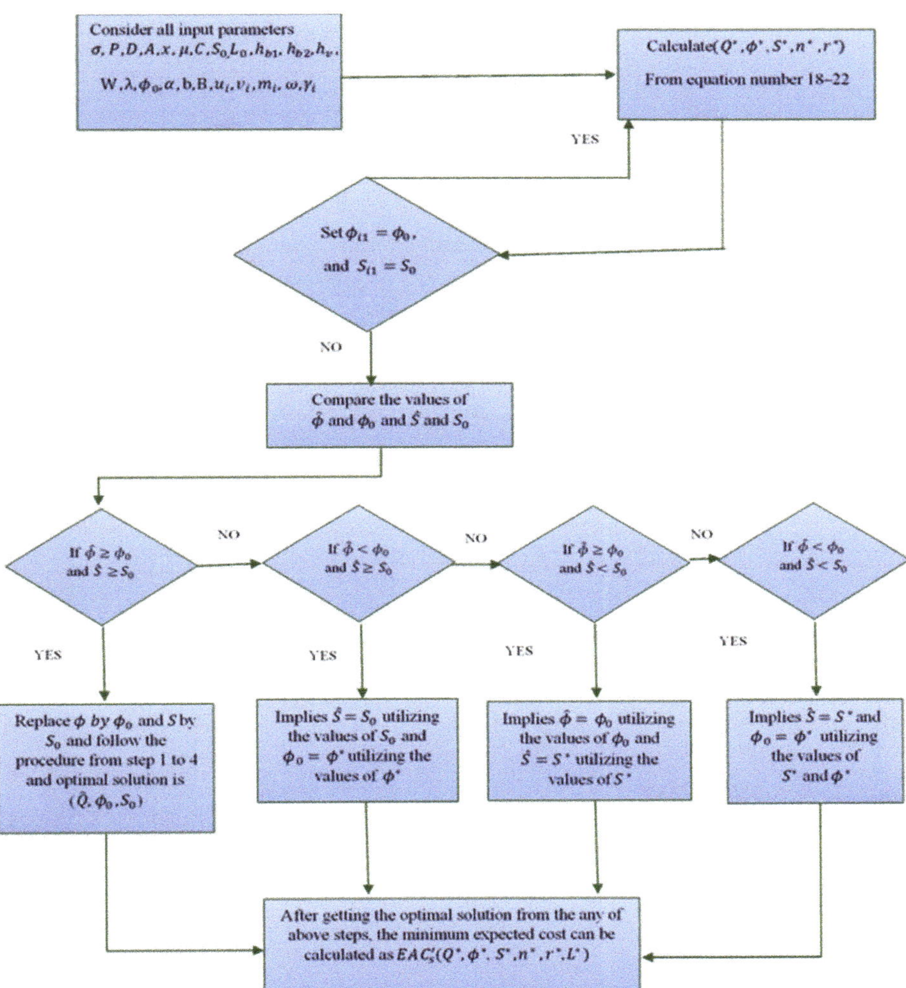

Figure 5. Flowchart of the algorithm.

Algorithm 1. Solution algorithm.

Step 1	Consider all input parameters, σ, P, D, A, x, μ, C, S_0, L_0, h_{b1}, h_{b2}, h_v, s, W, λ, ϕ_0, α, b, B, u_i, v_i, m_i, ω and γ_i.
Step 2	For each L_i, $i = 0, 1, 2, \ldots, n$, and execute Steps 2.1 to 2.2. Set $\phi_{i1} = \phi_0$, and $S_{i1} = S_0$.
Step 2.1	Substitute the value of ϕ_{i1} and S_{i1} in Equation (18) to evaluate Q_{i1}.
Step 2.2	Use the value of Q_{i1} to obtain ϕ_{i2}, and S_{i2} from Equations (19) and (20).
Step 2.3	Suppose $v = v + 1$, and repeat the above steps from 2.2 and 2.3 until no changes occur in the values of Q_i, ϕ_i, and S_i; then, denote the solution by $\hat{Q}$, $\hat{\phi}$ and $\hat{S}$.
Step 2.4	In this step, $\hat{\phi}$ and ϕ_0 and $\hat{S}$ and S_0 are compared. If $\hat{\phi} \le \phi_0$ and $\hat{S} \le S_0$, then the values obtained in Step 2 are the optimal solution for the given L_i;
Step 3	express this solution by $(Q^*, \phi^*, \text{ and } S^*)$ and then proceed to Step 4.
Step 3.1	If $\hat{\phi} \le \phi_0$ and $\hat{S} > S_0$, then set $S = S_0$ and use Step 2 to determine the $(\hat{Q}, \hat{\phi}, \hat{S})$ values from Equation (18) and Equation (20). If $\hat{\phi} \le \phi_0$, then the optimal solution for given L_i is $(Q^*, \phi^*, S^*) = (\hat{Q}, \hat{\phi}, S_0)$ and progress to Step 5. Otherwise, go to Step 3.3.
Step 3.2	If $\hat{\phi} > \phi_0$ and $\hat{S} \le S_0$, then set $\phi = \phi_0$ and use Step 2 to calculate new $(\hat{Q}, \hat{S})$ values from Equations (18) and (19). If $\hat{S} \le S_0$, then the optimal solution for given L_i is $(Q^*, \phi^*, S^*) = (\hat{Q}, \phi_0, \hat{S})$, and then proceed to Step 4. Otherwise, move to Step 3.4.
Step 3.3	$\hat{\phi} > \phi_0$ and $\hat{S} > S_0$; then set $\phi = \phi_0$ and $S = S_0$ and utilize Step 2 for calculating $\hat{Q}$ from Equation (18). Then, denote these values as $(Q^*, \phi^*, S^*) = (\hat{Q}, \phi_0, S_0)$ and move to Step 3.
Step 3.4	If all S_k, k = 1, 2, 3, $\ldots$, K are greater than or less than $\hat{S}$, chose the nearest S_k, and stop. Otherwise, calculate the $EAC_s'(Q^*, \phi^*, S^*, n, r, L_i)$ of the two nearest point using Equation (20) and select the S_k, with the least joint total cost. Express this solution by the vector $(Q^{**}, \phi^{**}, S^{**}, n, r, L_i)$.
Step 4	Choose the minimum cost for each L_i, $i = 0, 1, 2, \ldots, I$ $EAC_s'(Q^{**}, \phi^{**}, S^{**}, n, r, L_i) = EAC_s'(Q^{**}, \phi^{**}, S^{**}, n, r, L_i)$.
Step 5	Let $n = n + 1$, and repeat Steps 2 to 5 until satisfying an inequality,
Step 5.1	$EAC_s'(Q^{**}, \phi^{**}, S^{**}, n, r, L_s^{**})EAC_s'(Q^{**}, \phi^{**}, S^{**}, n+1, r, L_i)$.
Step 5.2	Choose the minimum cost for each n
Step 6	$EAC_s'(Q_{opt}, S_{opt}, r_{opt}, \phi_{opt}, n_{opt}, L_{opt}) = EAC_s'(Q^{**}, \phi^{**}, S^{**}, n, r, L_s^{**})$, and calculate the reorder point $y = r - \mu$.

3.3.2. Decentralized Analysis (Stackelberg Approach)

In the Stackelberg approach, the buyer and the vendor are considered as two different entities each trying to minimize their own total cost. Hence, one is considered as the leader to decide the optimal decision and impose on the follower.

The two possible cases in this situation are presented in the following part.

Case 1: Vendor as a leader and buyer as a follower

In the first case, the vendor is the leader and buyer is the follower. Therefore, the buyer decides his optimal decision first. The buyer determines his optimal order quantity, the lead time, and the reorder point. The aim of the vendor is the reduction of the setup cost, the number of shipments, and the percentage of defects after taking inputs from the buyer's optimal decisions.

The expected annual cost of the buyer with energy cost consideration is given in Equation (26):

$$
\begin{aligned}
EAC_b'(Q, r, L) = {} & \frac{D(A+A')}{Q} + \frac{DC(L)'}{Q} + \left(h_{b1} + h_{b1}'\right)\left[\phi Q - \frac{\phi(1+\phi)DQ}{2x}\right] + (h_{b2} \\
& + h_{b2}')\left[\frac{Q}{2} + \left(\frac{\sigma^2 L}{4(1-\lambda)Q} - (1-\lambda)Q\right) + \frac{\phi(1+\phi)DQ}{2x}\right] \\
& + (s + s')(1 + \phi) + D\left(\gamma_i + \gamma_i'\right),
\end{aligned}
\tag{26}
$$

and the expected annual cost of the vendor with energy consideration is given by Equation (27):

$$EAC_v'(S, \phi, n) = \quad \alpha b \ln \ln \left(\frac{\phi_0}{\phi}\right) + \alpha B \ln \ln \left(\frac{S_0}{S}\right) + \frac{D(S+S')}{nQ}$$
$$+ \frac{(h_v + h_v')Q}{2}[n(1 - D\rho) - 1 + 2D\rho] + (W + W')D\phi. \tag{27}$$

The analytical approach is applied to optimize the expected annual cost of the buyer. The partial derivative of the buyer's cost with respect to the order quantity is shown in Equation (28), and the partial derivative of lead time is shown in Equation (29):

$$\frac{\partial EAC_b'(Q,r,L)}{\partial Q} = \quad -\frac{DC(L)'}{Q^2} - \frac{(A+A')D}{Q^2} + \left[\phi - \frac{D\phi(1+\phi)}{2x}\right](h_{b1} + h_{b1}')$$
$$+ \left[-\frac{1}{2} + \frac{D\phi(1+\phi)}{2x} + \lambda - \frac{L\sigma^2}{4Q^2(1-\lambda)}\right](h_{b2} + h_{b2}'). \tag{28}$$

$$\frac{\partial EAC_b'(Q, r, L)}{\partial L} = \frac{\sigma^2(h_{b2} + h_{b2}')}{4Q(1 - \lambda)}. \tag{29}$$

The optimal values of the vendor are decided by considering the buyer's optimal decisions (Q, r, L) as inputs shown in Equations (30) and (31):

$$\frac{\partial EAC_v'(\phi, S, n)}{\partial \phi} = D(W + W') - \frac{b\alpha}{\phi}, \tag{30}$$

$$\frac{\partial EAC_v'(\phi, S, n)}{\partial S} = \frac{D}{nQ} - \frac{B\alpha}{(S + S')}. \tag{31}$$

Further, the second-order derivative of the buyer's ordering cost is shown in Equation (32):

$$\frac{\partial^2 EAC_v'(Q, r, L)}{\partial Q^2} = \frac{2C(L)'D}{Q^3} + \frac{2(A + A')D}{Q^3} + \frac{L\sigma^2 h_2}{2Q^3(1 - \lambda)}. \tag{32}$$

The second-order derivative of the vendor's decision variables are shown in Equations (33) and (34):

$$\frac{\partial^2 EAC_v'(\phi, S, n)}{\partial S^2} = \frac{B\alpha}{(S + S')^2}, \tag{33}$$

$$\frac{\partial^2 EAC_v'(\phi, S, n)}{\partial \phi^2} = \frac{b\alpha}{\phi^2}. \tag{34}$$

To find the optimal value of the order quantity, the partial derivatives of $\partial EAC_b'(Q, r, L)$ are equated to zero; thus, the optimal Q is shown in Equation (35):

$$Q'^* = \frac{\sqrt{x\left(-4D\left((A + A') + C(L)'(\lambda - 1) + L\sigma^2(h_{b2} + h_{b2}')\right)\right)}}{\sqrt{2}\sqrt{(\lambda - 1)\left(\phi(D - 2x + D\phi)\left(h_{b1} + h_{b1}'\right) - (D\phi(1 + \phi) + x(2\lambda - 1))(h_{b2} + h_{b2}')\right)}}. \tag{35}$$

The vendor will take the inputs from the buyer's optimal decision and compute his optimal decision, as shown in Equations (36) and (37):

$$\phi'^* = \frac{b\alpha}{D(W + W')}, \tag{36}$$

$$S'^* = \frac{BnQ\alpha}{D}. \tag{37}$$

Case 2: Buyer as a leader and vendor as a follower

In this case, the buyer is the leader and vendor is the follower. Therefore, the vendor optimizes his order quantity, the number of shipments, the reduction in setup cost, and the process quality improvement. The buyer will optimize the reorder point and the lead time based on the vendor optimal decision variables.

The expected annual cost of the vendor with energy consumption cost is shown in Equation (38):

$$EAC'_v(Q, n, S, \phi) = \alpha b \ln \ln \left(\frac{\phi_0}{\phi}\right) + \alpha B \ln \ln \left(\frac{S_0}{S}\right) + \frac{D(S+S')}{nQ}$$
$$+ \frac{(h_v + h'_v)Q}{2}[n(1 - D\rho) - 1 + 2D\rho] + (W + W')D\phi, \tag{38}$$

and the expected annual cost of the buyer with energy consumption cost is presented in Equation (39):

$$EAC'_b(r, L) = \frac{D(A+A')}{Q} + \frac{DC(L)'}{Q} + \left(h_{b1} + h'_{b1}\right)\left[\phi Q - \frac{\phi(1+\phi)DQ}{2x}\right]$$
$$+ (h_{b2} + h'_{b2})\left[\frac{Q}{2} + \left(\frac{\sigma^2 L}{4(1-\lambda)Q} - (1-\lambda)Q\right) + \frac{\phi(1+\phi)DQ}{2x}\right] + (s+s')D(1+\phi) + D\left(\gamma_i + \gamma'_i\right). \tag{39}$$

The expected annual cost of the vendor is optimized using an analytical method. The partial derivative of vendor costs with respect to the order quantity, the reduction in setup cost, and the defective percentage are shown in Equations (40)–(42):

$$\frac{\partial EAC'_v(Q, n, \phi, S)}{\partial Q} = -\frac{D(S+S')}{nQ^2} + \frac{1}{2}(2D\rho - 1 + n(1 - D\rho))(h_v + h'_v), \tag{40}$$

$$\frac{\partial EAC'_v(Q, n, \phi, S)}{\partial S} = \frac{D}{nQ} - \frac{B\alpha}{(S+S')}, \tag{41}$$

$$\frac{\partial EAC'_v(Q, n, \phi, S)}{\partial \phi} = D(W + W') - \frac{b\alpha}{\phi}. \tag{42}$$

Further, the second-order derivatives with respect to decision variable are shown in Equations (43) and (44):

$$\frac{\partial^2 EAC'_v(Q, n, \phi, S)}{\partial Q^2} = \frac{2D(S+S')}{nQ^3}, \tag{43}$$

$$\frac{\partial^2 EAC'_v(Q, n, \phi, S)}{\partial S^2} = \frac{B\alpha}{(S+S')^2}, \tag{44}$$

To obtain the optimal values of the lot size, the setup cost, and the process quality improvement, the partial derivatives of $EAC'_v(Q, n, r, L)$ are equated to zero. The optimal values are shown in Equations (45)–(47):

$$Q'^* = \frac{\sqrt{2D(S+S')}}{\sqrt{-n(1 - 2D\rho + n(-1 + D\rho))(h_v + h'_v)}}, \tag{45}$$

$$S'^* = \frac{BnQ\alpha}{D}, \tag{46}$$

$$\phi'^* = \frac{b\alpha}{D(W + W')}. \tag{47}$$

4. Numerical Examples

The following two numerical examples are considered to demonstrate the practical applicability of the proposed supply chain management model with relation to the electrical energy consumption. The data for the lead-time components shown in Table 3 are taken from Ouyang and Wu [13], but in their model, the cost of electrical energy consumption is not considered. However, in this numerical example, the lead time is reduced by considering an additional crashing cost and electrical energy consumption cost, as the first component cost reduction is minimum, and the energy consumed during this component is also minimum. Moving toward the second and third components, the cost increase with the energy consumption is presented in Table 3. The transportation cost structure presented in Table 4 (from Shin et al. [34]) is used in both numerical examples. However, in their model, they did not consider the cost of energy consumption with the transportation cost. In the given examples, the cost of electricity consumption is recognized, and the cost structure is shown in Table 4. The discount is offered for a certain transportation quantity; however, as the quantity increases, the consumption of energy also increases. Therefore, the cost of energy consumption increases in the transportation cost structure.

Table 3. Components of lead time.

Lead-Time Component	Normal Duration v_i (Days)	Minimum Duration u_i (Days)	Crashing Cost Per Unit m_i (\$/Days)	Cost of Electrical Energy Consumed m_i' (\$/Days)
1	20	6	0.3	0.1
2	20	6	1.0	0.2
3	16	9	4.5	0.5

Table 4. Transportation cost structure.

Quantity Range (units)	Unit Transportation Cost (\$/unit)	Unit Transportation Cost with Energy (\$/unit)
$0 \le Q < 200$	0.18	0.02
$200 \le Q < 400$	0.13	0.02
$400 \le Q < 600$	0.17	0.02
$600 \le Q$	0.14	0.3

Example 1. *The following parametric values (from Dey and Giri [6]) are used in the numerical analysis of the proposed model.* $D = 1000$ *units/year,* $\sigma = 7$, $\rho = 1/3200$, $A = \$49/setup$ $A' = \$1/setup$, $S_0 = \$400/setup$, $h_{b1} = \$5.8/unit/year$, $h_{b2} = \$9/unit/year$, $h_v = \$3.9/unit/year$, $s = \$0.22/unit$, $x = 2152$ *units/year,* $W = \$19/defective/unit$, $h_{b1}' = \$0.2/unit/year$, $h_{b2}' = \$1/unit/year$, $h_v' = \$0.1/unit/year$, $s' = \$0.0.03/unit$, $W' = \$1/defective$ $b = 400$, $B = 4000$, $\alpha = \$0.1/year$, $\phi_0 = 0.022$, $\lambda = 0.99$.*

The impact of the lead time on EAC_s with all the decision variables is shown in Table 5, and it is quite clear that when the lead time is three weeks, then the expected annual costs of the supply chain is minimum in comparison of four, six, and eight weeks based on the numerical example. From Table 5, it is also clear that when the lead time is minimal—i.e., the cost of energy consumption was minimal—that is the reason that the total cost is minimal. If the lead time is increased, during the waiting time, the energy is still consumed. Simultaneously, the effect of the changes on the related decision is also shown.

Table 5. Impact of lead time on EAC_s and other decision variables.

Lead-Time weeks	EAC_s \$/year	Q units	n lots	ϕ	S \$/year	r units
3 *	3295.45 *	176.16 *	2 *	0.00183 *	140.93 *	30.2 *
4	3380.87	124.52	5	0.00187	249.05	49.10
6	3538.68	119.34	6	0.00188	286.42	71.05
8	3317.19	178.04	2	0.00183	242.43	64.26

* The bold numbers are the optimum values obtained.

In Table 6, the EAC_s is calculated when the investment is made for the production quality improvement and without any additional initial capital investment. Based on the numerical study, it can be concluded that there was a large reduction in the total supply chain cost with investment in process quality improvement and setup cost reduction. The initial investment for setup cost reduction and quality improvement also affect the energy consumption of the supply chain, which is shown in Table 6, and lowers the overall supply chain costs.

Table 6. Comparison of EAC_s with and without investment in the reduction of the setup cost and quality enhancing.

EAC_s with Quality Improvement, and Setup Cost Reduction	EAC_s	EAC_s When No Investments are Made for Process Quality Improvement and Setup Cost Reduction	EAC_s
(Q, n, S, ϕ, r, L)	\$/year	(Q, n, r, L)	
(176.16, 2, 0.0018, 140.93, 30.2, 3)	3295.45	(66.99, 23, 65.18, 3)	5628.44

Table 7 shows the comparative EAC_s results for the improvement in the process quality and the reduction in the setup cost with no investment for quality improvement. Based on the numerical experiment, the investment made for quality improvement reduces the supply chain cost by improving the process quality. In Table 7, investments are not made for quality improvement, which means that the consumption of energy is relatively high, but the overall supply chain cost is still less than the cost without initial investments for quality improvement and setup cost reduction. The reason for total cost reduction is savings from the setup cost reduction through investment in advanced technology in the form of machines with less energy consumption.

Table 7. Comparison of EAC_s with and without investment in quality improvement.

EAC_s with Quality Improvement, and Setup Cost Reduction	EAC_s	EAC_s with No Investment for Process Quality Improvement	EAC_s
(Q, n, S, ϕ, r, L)	\$/year	(Q, n, r, S, L)	\$/year
(176.16, 2, 0.00183, 140.93, 30.2, 3)	3295.45	(116.82, 7, 41.29, 327.09, 3)	4154.63

In Table 8, the comparison of EAC_s results are shown, when investments are made for process quality improvement and setup cost reduction, with a fixed setup cost. From the numerical study, it is clear that a huge amount of money can be saved by the incorporation initial investment for the setup cost. The investment in setup cost reduction is very effective in total cost reduction for this model, because it gives the opportunity for to vendor to use advanced technologies in production that consume energy effectively. Moreover, from Table 8, if initial investments are not made for quality improvement, the overall cost is increased. The reason behind that is the wastage due to poor quality and more energy consumption due to reproduction and by holding the defective items.

Table 8. EAC_s with investment in setup cost reduction and no investment for setup cost.

EAC_s with Quality Improvement, and Setup Cost Reduction	EAC_s	EAC_s with No Investment for Setup Cost Reduction	EAC_s
(Q, n, S, ϕ, r, L)	\$/year	(Q, n, r, ϕ, L)	\$/year
(176.16, 2, 0.0018, 140.93, 30, 3)	3295.45	(89.08, 13, 51, 0.0019, 3)	4368.60

Tables 5–8 present the results of the centralized supply chain. Table 9 shows the decentralized supply chain optimal values, and based on the numerical experiment, it is clear that the centralized system is better with low cost. Furthermore, in the decentralized supply chain context, when the buyer

is the leader, the total cost of the supply chain is less compared to the case when the vendor is the leader by keeping same cost parameters.

Table 9. Comparative results for the centralized and decentralized supply chain management policies.

Leader	Follower	Q (units)	n	S ($/setup)	ϕ	r (units)	Buyer Cost ($/year)	Vendor Cost ($/year)	Total Cost ($/year)
Vendor	Buyer	45.07	7	126.20	0.002	50.86	3308.26	1402.4	4710.73
Buyer	Vendor	158.41	1	63.36	0.002	19.43	1894.17	1577.06	3471.23
Centralized supply chain		176.16	2	140.93	0.0018	30.20	N/A	N/A	3295.45

"N/A" indicates not applicable.

The impact of the decision variables on the expected annual cost is shown graphically in the figures below. Based on the data from numerical Example 1, and from Figure 6, it is quite clear that the total cost of the supply chain is optimum when the buyer orders 176 items from the vendor; accordingly, any change in the order quantity by keeping the other decision variables the same, increases the expected annual cost of the supply chain. The setup cost relationship is shown in Figure 7, which shows that when the setup cost is $140, then the optimum supply chain costs are obtained, and any further variation increases the overall cost.

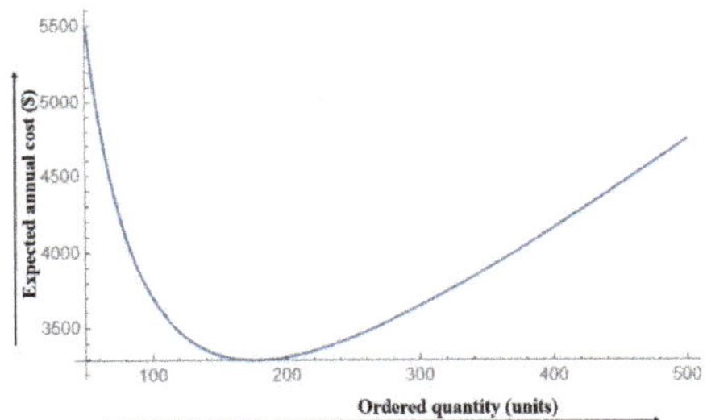

Figure 6. Order quantity versus EAC_s.

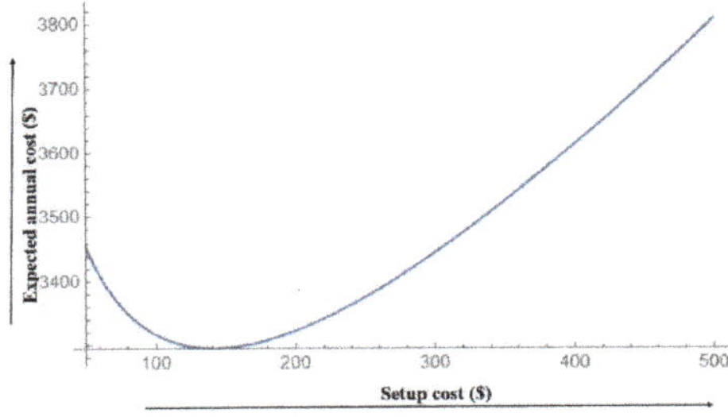

Figure 7. Setup cost versus EAC_s.

Figure 8 shows the relation between the out-of-control probability and the expected annual cost; the expected annual cost is optimum when the defective percentage is 0.0018, and changing the percentage of defective items increases the total supply chain cost. Similarly, from Figure 9, based on the data in numerical Example 1, the optimum number of shipments is 2, and any variation in the number of shipments—keeping the other parameters the same—will increase the expected annual cost of the supply chain.

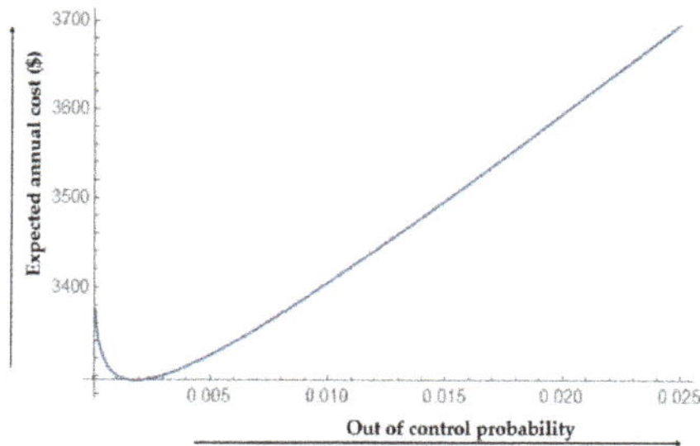

Figure 8. Out-of-control probability versus EAC_s.

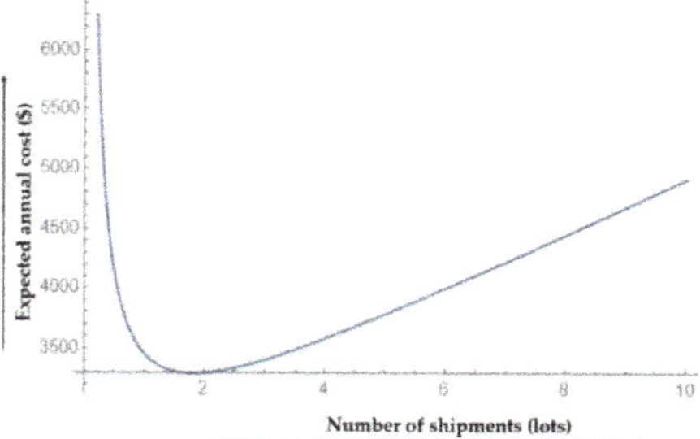

Figure 9. Number of shipments versus EAC_s.

Example 2. *The following parametric values (from Salameh and Jaber [3]) were used in the numerical analysis.* $D = 50000$ *units/year,* $\sigma = 200$, $\rho = 1/160000$, $A = \$98/setup$ $A' = \$2/setup$, $S_0 = \$300/setup$, $h_{b1} = \$2.8/unit/year$, $h_{b2} = \$4.7/unit/year$, $h_v = \$1.9/unit/year$, $s = \$0.45/unit$, $x = 175200$ *units/year,* $W = \$29/defective/unit$, $h'_{b1} = \$0.2/unit/year$, $h'_{b2} = \$0.3/unit/year$, $h'_v = \$0.1/unit/year$, $s' = \$0.0.05/unit$, $W' = \$1/defective$, $b = 400$, $B = 4000$, $\alpha = \$0.1/year$, $\phi_0 = 0.022$, *and* $\lambda = 0.99$.

Table 10 shows the impact of the lead time on the other decision variables and on the total cost. Based on the data shown in numerical Example 2, it is clear that the minimum cost is achieved when the lead time is four weeks. If the lead time increases from four weeks, it ultimately increases the total cost of the supply chain. In that case, more energy is consumed due to the increased lead time. Therefore, the lead crashing cost is used with energy consumption cost to reduce the lead time and energy consumption in this model.

Table 10. Impact of lead time on EAC_s and decision variables.

L weeks	EAC_s $/year	Q units	n	ϕ	S $/setup	r units
3	64105.46	1242.43	19	0.000026	188.85	2391.19
4 *	56424.93 *	1802.36 *	9 *	0.000026 *	129.77 *	2190.29 *
6	147017	304.46	14	0.000026	30.09	19692.9
8	56622	3095.62	4	0.000026	99.05	2542.33

* The bold numbers are the optimum values obtained.

In Table 11, the comparison of the optimal costs of the proposed supply chain model and the optimal cost without investment in the process quality improvement and the setup cost reduction is provided. The reduction in the expected annual cost of the supply chain clearly describes the importance of initial capital investments for process quality improvement and setup cost reduction. The total cost of the supply chain reduces by the efforts made through initial capital investments. These investments are utilized for purchasing the technological upgraded machinery, which consumed less energy and improved production. Thus, these investments helped reduce the total supply chain costs.

Table 11. Comparison of EAC_s with reduction in setup cost and quality improvement, and fixed setup cost and quality.

EAC_s with Quality Improvement and Setup Cost Reduction	EAC_s	EAC_s without Quality Improvement and Setup Cost Reduction	EAC_s
(Q, n, S, ϕ, r, L)	\$/year	(Q, n, r, L)	\$/year
$(1802.36, 9, 129.77, 0.000026, 2190.29, 4)$	56424.93	$(350.34, 307, 11{,}402.8, 4)$	$205{,}889$

The effect of investment in quality improvement on total cost is depicted in Table 12; further, it is quite clear after comparing it with the model in which no investment was made for quality, which increases the expected annual cost. From Table 12, it is quite clear that the initial capital invested for quality improvement is utilized in innovative ways to reduce the defective item percentage in the model. The total cost is decreased by quality improvement because of the production of less defective items, which consumed less energy.

Table 12. Comparison of EAC_s with and without investment in quality improvement.

EAC_s with Quality Improvement and Setup Cost Reduction	EAC_s	EAC_s without Quality Improvement	EAC_s
(Q, n, S, ϕ, r, L)	\$/year	(Q, n, S, r, L)	\$/year
$(1802.3, 9, 129.7, 0.000026, 2190.29, 4)$	56424.93	$(1967.14, 7, 110.16, 2002.74, 4)$	$84{,}257.1$

In Table 13, the comparison is shown in which investments are made in the setup cost, with no initial capital investment. Consequently, the initial capital investment for the setup cost reduces the total supply chain costs with lower energy consumption in the supply chain model.

Table 13. EAC_s with investment in setup cost reduction and quality with no investments for both.

EAC_s with Quality Improvement and Setup Cost Reduction	EAC_s	EAC_s without Investment in Setup Cost Reduction	EAC_s
(Q, n, S, ϕ, r, L)	\$/year	(Q, n, ϕ, r, L)	\$/year
$(1802.36, 9, 129.7, 0.000026, 2190.29, 4)$	$56{,}424.93$	$(428.99, 204, 0.000026, 9308.7, 4)$	$148{,}420$

The effect on total cost with the decision variable is graphically shown in the following figures. Based on the data from numerical Example 2, Figure 10 shows that the optimal order quantity Q is 1802 units, and that further change in the order quantity—keeping the other parameters the same—increases the expected annual cost. Figure 11 depicts that the optimal setup cost is \$129 for the model based on the data from numerical Example 2; any change in the setup cost without changing the other parameters increases the expected annual cost of the supply chain.

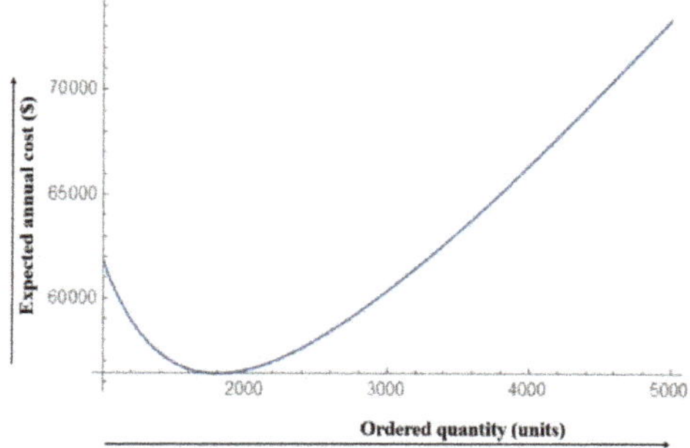

Figure 10. Order quantity versus EAC_s.

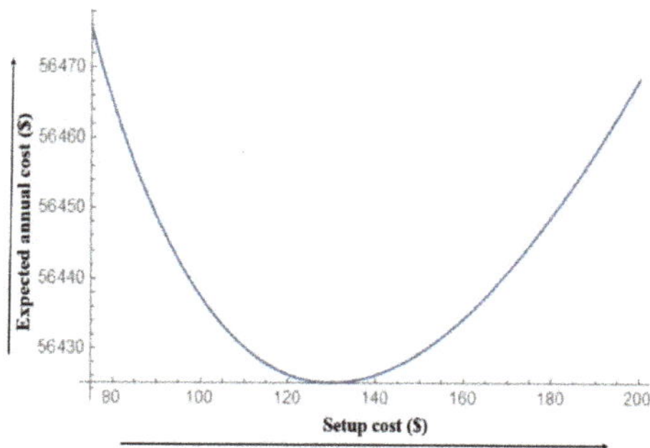

Figure 11. Setup cost versus EAC_s.

The effect of a change in the out-of-control probability is shown in Figure 12 for the annual expected cost. From Figure 13, based on the data from numerical Example 2, it is quite clear that the optimal number of shipments is two, and increasing or decreasing the number of shipments increases the annual expected cost.

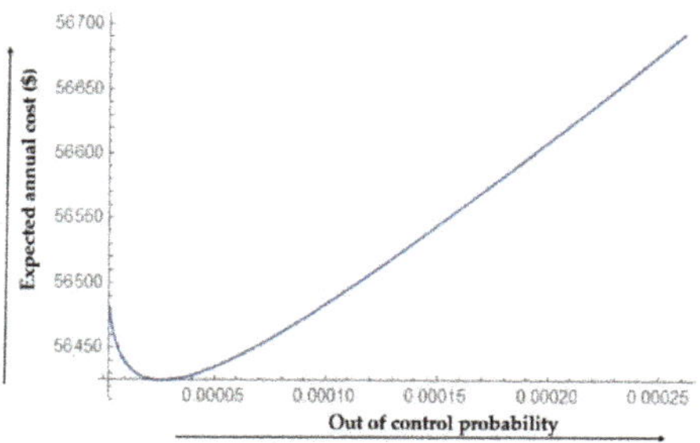

Figure 12. Out-of-control probability versus EAC_s.

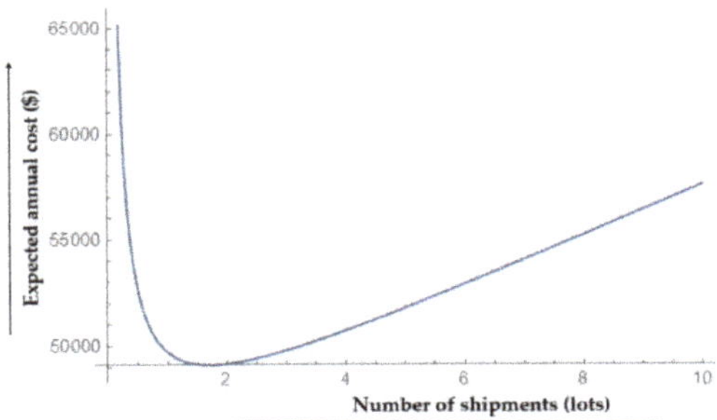

Figure 13. Number of shipments versus EAC_s.

5. Sensitivity Analysis

The sensitivity analysis is done for all the key parameters of the optimal values for Example 1, and is shown in Table 14. The values of the key parameters are varied from −50% to +50%, respectively, and the remaining parameters are unchanged. The effect of these parameters on the decision variables is calculated. The percentage variation in the optimal cost illustrates the following crucial features.

- If the investment in the quality is reduced by 50%, a small increase in the expected annual cost is observed, which means that the quality of the product is affected, and the extra energy is consumed through reproducing and keeping defective items. Similarly, if the investment in the quality is reduced by 25%, then a significant increase in the total cost is observed due to extra energy consumption on the defective items. The capital cost is more sensitive to the increase in investment than to its reduction.

- The negative change for the capital investment in setup cost reduces the total cost, but if the setup cost increases, the total cost also increases. The capital investment for the setup cost reduces the optimal order quantity.

- The fractional cost for the capital investment has an asymmetrical effect on the total cost, and it is considered as a sensitive parameter in this model. If the capital investments are reduced by 25%, the cost increases by 25%.

- The holding cost of the buyer for defective items is less sensitive than the total cost, because the amount of the defective products is less.

- The holding cost of the buyer's non-defective items is the most sensitive parameter. If the holding decreases by 50%, the total cost reduces by 7%. Further, a 50% increase in the holding cost leads to an increase in the total cost by 16%. The inventory of the buyer's non-defective items is more in quantity in this supply chain model; this is the reason behind the increase in total cost, and it consumes more energy.

- The variation in the holding cost of the vendor affects the total cost. If the holding cost decreases, the total cost reduces. Similarly, by increasing the holding cost, the total cost increases. A huge amount of energy is consumed while holding the products, and it also increases the holding cost, which ultimately increases the total cost of the supply chain model.

- The screening cost with energy consumption cost has a symmetrical effect on the total cost differently from the other parameters, which has an asymmetric impact on the entire cost of the supply chain, and it has no effect on the other decision variables. The screening cost includes the energy consumption cost during the screening and the variation observed for −50% and +50% is around 4% of the total cost of the supply chain.

- If the cost for the defective unit replacement with the energy consumption cost is decreased by 50%, then the total cost is reduced by a small percentage; however, if the decrement is 25%, then the total cost is increased by 10%. There is a small effect of positive percentage increase on the defective unit replacement cost.

Table 14. Sensitivity analysis for key parameters.

Input Parameters	Percentage (%) Changes in Value	Percentage Changes in EAC_s
b	−50	0.26
	−25	1.20
	+25	2.78
	+50	1.45
B	−50	−11.35
	−25	−3.65
	+25	14.51
	+50	4.72
α	−50	−6.38
	−25	25.55
	+25	6.84
	+50	6.17
h_{b1}	−50	−0.030
	−25	2.02
	+25	0.01
	+50	0.02
h_{b2}	−50	−6.45
	−25	−8.36
	+25	7.95
	+50	15.63
h_v	−50	−1.86
	−25	−2.73
	+25	1.51
	+50	2.44
s	−50	−3.80
	−25	−1.90
	+25	1.90
	+50	3.80
W	−50	−0.74
	−25	10.79
	+25	0.25
	+50	0.45

6. Managerial Insights

Based on this study, the following recommendations are suggested for decision makers dealing with imperfect production in supply chain management:

- The major recommendation from this study for decision makers is the capital investment for process quality improvement. The investment is made for bringing in new technology, which uses optimum energy consumption; simultaneously, it reduces the probability that the system will move to an out-of-control state. It is proven from the numerical experiment that the investment reduces the total cost of the supply chain with minimum energy consumption.

- Another important insight for the present study from the decision maker's point of view is the initial capital investment for setup cost reduction. This model suggests that decision makers ought to invest more initially to get the latest machinery and equipment, which can be used to reduce the whole supply chain cost with minimum energy consumption, as proven numerically.
- Due to the controllable lead time, if the buyer must make the decision related to lead-time reduction, by facing an additional crashing cost, he can reduce the lead time.
- In this model, the stockout costs are replaced with the service-level constraint, which helps managers overcome the difficult situations that occur due to the shortages, as well as the, enormous amount of energy consumed during the supply chain process. To avoid such situations, this model suggests the optimum level of inventory to be satisfied from the available stock to reduce energy consumption.
- In many cases, the objective of the managers is to minimize the cost by optimizing the decision variables. In this model, the energy costs are considered to elaborate the importance of the quantity of the consumed energy in the system, which is not considered in available supply chain models. If the cost incurred on the energy consumption is known to the managers, then the optimized level of energy consumption can be decided.

7. Conclusions

A two-echelon supply chain management model was developed in this study considering the effect of imperfect production, the inspection process, and energy consumption. The energy consumption related to the costs was not considered in previous research studies, and was proven to have numerous effects on the overall cost of the supply chain. As a result of introducing the energy consumption cost during the supply chain operation, a huge amount of money is saved. Furthermore, in this model, to reduce the shortages due to defective item production, the vendor sends an extra amount of quantity equal to the defective percentage to the buyer, and the collection of this amount takes place in the next batch. A lemma was constructed to show that the model attends a global optimal solution. This model encourages the managers to make an initial investment in order to improve the process quality and reduce the setup cost, which reduces the overall supply chain costs with enough energy consumption. Based on the analysis of the model, it was proven numerically that the centralized system is the most efficient compared to decentralized cases. However, in decentralized cases, the optimum cost is low when the buyer is a leader and the vendor is a follower. Based on numerical Example 1, the expected annual cost is optimum when the lead time is three weeks, and from numerical Example 2, the optimum expected annual cost is obtained when the lead time is four weeks. The effect of the initial investment made for quality improvement and setup cost reduction has a significant effect on the expected annual cost reduction and minimum energy consumption, as concluded from numerical examples. Furthermore, the graphical analysis is performed to show the impact of the variation in the lot size, the number of shipments, the setup cost, and the process quality by considering energy-related costs. The sensitivity analysis was performed to validate the proposed model. The proposed model in this paper can be extended in several aspects by including the green supply chain management with variable demand, multi-item production, a multi-buyer supply chain, and a region-based distribution network.

Author Contributions: Conceptualization, I.K. and B.S.; methodology, investigation, formal analysis, writing—original draft preparation, I.K.; conceptualization, resources, validation, writing—review and editing, visualization, supervision, B.S.; data curation, software, I.K. and J.J.; data curation and software, H.L.

Funding: There is no funding for the research paper.

Acknowledgments: The authors are thankful to three anonymous reviewers for their valuable comments to revise the older version of the paper.

Appendix A

Lemma 1. *If the Hessian matrix for EAC_s is positive definite at the optimal values (Q^*, S^*, ϕ^*) then EAC_s contains the global minimum at the optimum solution (Q^*, S^*, ϕ^*).*

Proof. To show that the Hessian matrix for EAC_s is always positive definite, all minors must be positive definite, and the corresponding principal minors at the optimal values need to be:

$$det H_{11} = det\left(\frac{\partial^2 EAC_s}{\partial Q^2}\right) = \frac{2\big(C(L) + C(L)'\big)D}{Q^3} + \frac{2(A + A')D}{Q^3} + \frac{2D(S + S')}{nQ^3} + \frac{(h_{b2} + h'_{b2})L\sigma^2}{2Q^3(1 - \lambda)} > 0$$

$$det H_{22} = det\left(\frac{\partial^2 EAC_s}{\partial Q^2} \quad \frac{\partial^2 EAC_s}{\partial Q\partial S} \quad \frac{\partial^2 EAC_s}{\partial S\partial Q} \quad \frac{\partial^2 EAC_s}{\partial S^2}\right)$$

$$= \frac{D^2}{n^2 Q^4(\lambda - 1)} - \frac{2BC(L)D\alpha}{Q^3 S^2(\lambda - 1)} - \frac{2ABD\alpha}{Q^3 S^2(\lambda - 1)} - \frac{2BD\alpha}{nQ^3 S(\lambda - 1)} - \frac{D^2\lambda}{n^2 Q^4(\lambda - 1)} + \frac{2BC(L)D\alpha\lambda}{Q^3 S^2(\lambda - 1)} + \frac{2ABD\alpha\lambda}{Q^3 S^2(\lambda - 1)} + \frac{2BD\alpha\lambda}{nQ^3 S(\lambda - 1)}$$
$$- \frac{Bh_{b2}L\alpha\sigma^2}{2Q^3 S^2(\lambda - 1)} > 0$$

$$H_{33} = det\left(\frac{\partial^2 EAC_s}{\partial Q^2} \quad \frac{\partial^2 EAC_s}{\partial Q\partial S} \quad \frac{\partial^2 EAC_s}{\partial Q\partial \phi} \quad \frac{\partial^2 EAC_s}{\partial S\partial Q} \quad \frac{\partial^2 EAC_s}{\partial S^2} \quad \frac{\partial^2 EAC_s}{\partial S\partial \phi} \quad \frac{\partial^2 EAC_s}{\partial \phi\partial Q} \quad \frac{\partial^2 EAC_s}{\partial \phi\partial S} \quad \frac{\partial^2 EAC_s}{\partial \phi^2}\right)$$

$$= \frac{1}{4n^2 Q^5 S^2 x^2(\lambda - 1)\phi^2}\left(-Bn^2 Q^5 \alpha(\lambda - 1)\phi^2(-Dh_{b1} + Dh_{b2} + 2h_{b1}x - 2Dh_{b1}\phi + 2Dh_{b2}\phi)^2\right.$$
$$+ 2(-2D^2 Q S^2 x(\lambda - 1) + BnQ^2 x\alpha(-4ADn - 4C(L)Dn - 4DS + 4ADn\lambda$$
$$\left. + 4CDn\lambda + 4DS\lambda - h_{b2}Ln\sigma^2))\big(bx\alpha - Dh_{b1}Q\phi^2 + djQ\phi^2\big)\right) > 0$$

It is proved that when all the principal minors are positive definite, EAC_s contains the optimal solution (Q^*, S^*, ϕ^*). Hence, the EAC_s contains the global minimum at the optimal decision variable under the effect of optimum energy consumption. $\square$

References

1. Sarkar, M.; Sarkar, B. Optimization of safety stock under controllable production rate and energy consumption in an automated smart production management. *Energies* **2019**, *12*, 2059. [CrossRef]
2. Sarkar, B.; Saren, S.; Sinha, D.; Hur, S. Effect of unequal lot sizes, variable setup cost, and carbon emission cost in a supply chain model. *Math Probl. Eng.* **2015**, *2015*, 469486. [CrossRef]
3. Salameh, M.; Jaber, M. Economic production quantity model for items with imperfect quality. *Int. J. Prod. Econ.* **2000**, *64*, 59–64. [CrossRef]
4. Ouyang, L.Y.; Chen, C.K.; Chang, H.C. Quality improvement, setup cost and lead-time reductions in lot size reorder point models with an imperfect production process. *Comput. Oper. Res.* **2002**, *29*, 1701–1717. [CrossRef]
5. Ben-Daya, M.; Hariga, M. Integrated single vendor single buyer model with stochastic demand and variable lead time. *Int. J. Prod. Econ.* **2004**, *92*, 75–80. [CrossRef]
6. Dey, O.; Giri, B. Optimal vendor investment for reducing defect rate in a vendor–buyer integrated system with imperfect production process. *Int. J. Prod. Econ.* **2014**, *155*, 222–228. [CrossRef]
7. Khan, M.; Jaber, M.; Guiffrida, A.; Zolfaghari, S. A review of the extensions of a modified EOQ model for imperfect quality items. *Int. J. Prod. Econ.* **2011**, *132*, 1–12. [CrossRef]
8. Chung, K.J. The economic production quantity with rework process in supply chain management. *Comput. Math. Appl.* **2011**, *62*, 2547–2550. [CrossRef]
9. Tayyab, M.; Sarkar, B. Optimal batch quantity in a cleaner multi-stage lean production system with random defective rate. *J. Clean. Prod.* **2016**, *139*, 922–934. [CrossRef]
10. Cárdenas-Barrón, L.E.; Taleizadeh, A.A.; Treviño-Garza, G. An improved solution to replenishment lot size problem with discontinuous issuing policy and rework, and the multi-delivery policy into economic production lot size problem with partial rework. *Expert Syst. Appl.* **2012**, *39*, 13540–13546. [CrossRef]
11. Scarf, H. A min-max solution of an inventory problem. In *Studies in the Mathematical Theory of Inventory and Production*; Stanford University Press: Redwood City, CA, USA, 1958.
12. Gallego, G.; Moon, I. The distribution free newsboy problem: Review and extensions. *J. Oper. Res. Soc.* **1993**, *44*, 825–834. [CrossRef]

13. Ouyang, L.Y.; Wu, K.S. Mixture inventory model involving variable lead time with a service level constraint. *Comput. Oper. Res.* **1997**, *24*, 875–882. [CrossRef]

14. Moon, I.; Choi, S. The distribution free continuous review inventory system with a service level constraint. *Comput. Ind. Eng.* **1994**, *27*, 209–212. [CrossRef]

15. Tajbakhsh, M.M. On the distribution free continuous-review inventory model with a service level constraint. *Comput. Ind. Eng.* **2010**, *59*, 1022–1024. [CrossRef]

16. Ma, W.M.; Qiu, B.B. Distribution-free continuous review inventory model with controllable lead time and setup cost in the presence of a service level constraint. *Math. Probl. Eng.* **2012**, *2012*, 867847. [CrossRef]

17. Harris, F.W. How many parts to make at once. In *Factory, The Magazine of Management*; University of Michigan: Arbor, MI, USA, 1913.

18. Cárdenas-Barrón, L.E. Observation on:" Economic production quantity model for items with imperfect quality" [Int. J. Production Economics 64 (2000) 59-64]. *Int. J. Prod. Econ.* **2000**, *67*, 201.

19. Ciavotta, M.; Meloni, C.; Pranzo, M. Minimising general setup costs in a two-stage production system. *Int. J. Prod. Res.* **2013**, *51*, 2268–2280. [CrossRef]

20. Dellino, G.; Kleijnen, J.P.; Meloni, C. Robust optimization in simulation: Taguchi and response surface methodology. *Int. J. Prod. Econ.* **2010**, *125*, 52–59. [CrossRef]

21. Goyal, S.K.; Cárdenas-Barrón, L.E. Note on: Economic production quantity model for items with imperfect quality–A practical approach. *Int. J. Prod. Econ.* **2002**, *77*, 85–87. [CrossRef]

22. Huang, C.K. An optimal policy for a single-vendor single-buyer integrated production–inventory problem with process unreliability consideration. *Int. J. Prod. Econ.* **2004**, *91*, 91–98. [CrossRef]

23. Kim, M.S.; Kim, J.S.; Sarkar, B.; Sarkar, M.; Iqbal, M.W. An improved way to calculate imperfect items during long-run production in an integrated inventory model with backorders. *J. Manuf. Syst.* **2018**, *47*, 153–167. [CrossRef]

24. Sarkar, B. Supply chain coordination with variable backorder, inspections, and discount policy for fixed lifetime products. *Math Probl. Eng.* **2016**, *2016*, 6318737. [CrossRef]

25. Tang, L.; Che, P.; Liu, J. A stochastic production planning problem with nonlinear cost. *Comput. Oper. Res.* **2012**, *39*, 1977–1987. [CrossRef]

26. Gahm, C.; Denz, F.; Dirr, M.; Tuma, A. Energy-efficient scheduling in manufacturing companies: A review and research framework. *Ejor* **2016**, *248*, 744–757. [CrossRef]

27. Keller, F.; Reinhart, G. Energy Supply Orientation in Production Planning Systems. *Procedia Cirp* **2016**, *40*, 244–249. [CrossRef]

28. Du, Y.; Hu, G.; Xiang, S.; Zhang, K.; Liu, H.; Guo, F. Estimation of the diesel particulate filter soot load based on an equivalent circuit model. *Energies* **2018**, *11*, 472. [CrossRef]

29. Todde, G.; Murgia, L.; Caria, M.; Pazzona, A. A comprehensive energy analysis and related carbon footprint of dairy farms, Part 2: Investigation and modeling of indirect energy requirements. *Energies* **2018**, *11*, 463. [CrossRef]

30. Tomić, T.; Schneider, D.R. The role of energy from waste in circular economy and closing the loop concept–Energy analysis approach. *Renew. Sustain. Energy Rev.* **2018**, *98*, 268–287. [CrossRef]

31. Haraldsson, J.; Johansson, M.T. Review of measures for improved energy efficiency in production-related processes in the aluminium industry–From electrolysis to recycling. *Renew. Sustain. Energy Rev.* **2018**, *93*, 525–548. [CrossRef]

32. Ouyang, L.Y.; Chang, H.C. Impact of investing in quality improvement on (Q, r, L) model involving the imperfect production process. *Prod. Plan. Cont.* **2000**, *11*, 598–607. [CrossRef]

33. Porteus, E.L. Optimal lot sizing, process quality improvement and setup cost reduction. *Oper. Res.* **1986**, *34*, 137–144. [CrossRef]

34. Shin, D.; Guchhait, R.; Sarkar, B.; Mittal, M. Controllable lead time, service level constraint, and transportation discounts in a continuous review inventory model. *Rairo-Oper. Res.* **2016**, *50*, 921–934. [CrossRef]

35. Braglia, M.; Castellano, D.; Song, D. Distribution-free approach for stochastic Joint-Replenishment Problem with backorders-lost sales mixtures, and controllable major ordering cost and lead times. *Comput. Oper. Res.* **2017**, *79*, 161–173. [CrossRef]

36. Tsao, Y.C.; Lu, J.C. A supply chain network design considering transportation cost discounts. *Trans. Res. Part E Log. Rev.* **2012**, *48*, 401–414. [CrossRef]

37. Ouyang, L.Y.; Chen, L.Y.; Yang, C.T. Impacts of collaborative investment and inspection policies on the integrated inventory model with defective items. *Int. J. Prod. Res.* **2013**, *51*, 5789–5802. [CrossRef]
38. Jha, J.; Shanker, K. An integrated inventory problem with transportation in a divergent supply chain under service level constraint. *J. Manuf. Syst.* **2014**, *33*, 462–475. [CrossRef]
39. Priyan, S.; Uthayakumar, R. Trade credit financing in the vendor–buyer inventory system with ordering cost reduction, transportation cost and backorder price discount when the received quantity is uncertain. *J. Manuf. Syst.* **2014**, *33*, 654–674. [CrossRef]
40. Gutgutia, A.; Jha, J. A closed-form solution for the distribution free continuous review integrated inventory model. *Oper. Res.* **2018**, *18*, 159–186. [CrossRef]
41. Hadley, G.; Whitin, T.M. *Analysis of Inventory Systems: G. Hadley, TM Whitin*; Prentice-Hall: Upper Saddle River, NJ, USA, 1963.
42. Liao, C.J.; Shyu, C.H. An analytical determination of lead time with normal demand. *Int. J. Oper. Prod. Manag.* **1991**, *11*, 72–78. [CrossRef]

MDPI
St. Alban-Anlage 66
4052 Basel
Switzerland
Tel. +41 61 683 77 34
Fax +41 61 302 89 18
www.mdpi.com

Energies Editorial Office
E-mail: energies@mdpi.com
www.mdpi.com/journal/energies